DISCRETE RANDOM SIGNAL PROCESSING and FILTERING PRIMER with MATLAB®

THE ELECTRICAL ENGINEERING
AND APPLIED SIGNAL PROCESSING SERIES
Edited by Alexander D. Poularikas

The Transform and Data Compression Handbook
K.R. Rao and P.C. Yip

Handbook of Antennas in Wireless Communications
Lal Chand Godara

Handbook of Neural Network Signal Processing
Yu Hen Hu and Jenq-Neng Hwang

Optical and Wireless Communications: Next Generation Networks
Matthew N.O. Sadiku

Noise Reduction in Speech Applications
Gillian M. Davis

Signal Processing Noise
Vyacheslav P. Tuzlukov

Digital Signal Processing with Examples in MATLAB®
Samuel Stearns

Applications in Time-Frequency Signal Processing
Antonia Papandreou-Suppappola

The Digital Color Imaging Handbook
Gaurav Sharma

Pattern Recognition in Speech and Language Processing
Wu Chou and Biing-Hwang Juang

Propagation Handbook for Wireless Communication System Design
Robert K. Crane

Nonlinear Signal and Image Processing: Theory, Methods, and Applications
Kenneth E. Barner and Gonzalo R. Arce

Smart Antennas
Lal Chand Godara

Mobile Internet: Enabling Technologies and Services
Apostolis K. Salkintzis and Alexander D. Poularikas

Soft Computing with MATLAB®
Ali Zilouchian

Signal and Image Processing in Navigational Systems
Vyacheslav P. Tuzlukov

Medical Image Analysis Methods
Lena Costaridou

MIMO System Technology for Wireless Communications
George Tsoulos

Signals and Systems Primer with MATLAB®
Alexander D. Poularikas

THE ELECTRICAL ENGINEERING
AND APPLIED SIGNAL PROCESSING SERIES

DISCRETE RANDOM SIGNAL PROCESSING and FILTERING PRIMER with MATLAB®

Alexander D. Poularikas

CRC Press
Taylor & Francis Group
Boca Raton   London   New York

CRC Press is an imprint of the
Taylor & Francis Group, an **informa** business

CRC Press
Taylor & Francis Group
6000 Broken Sound Parkway NW, Suite 300
Boca Raton, FL 33487-2742

© 2009 by Taylor & Francis Group, LLC
CRC Press is an imprint of Taylor & Francis Group, an Informa business

No claim to original U.S. Government works
Printed in the United States of America on acid-free paper
10 9 8 7 6 5 4 3 2 1

International Standard Book Number-13: 978-1-4200-8933-2 (Hardcover)

Library of Congress Cataloging-in-Publication Data

Poularikas, Alexander D., 1933-
  Discrete random signal processing and filtering primer with MATLAB /
Alexander D. Poularikas.
      p. cm.
   Includes bibliographical references and index.
   ISBN 978-1-4200-8933-2 (alk. paper)
   1. Electric filters. 2. MATLAB. 3. Signal processing. I. Title.

TK7872.F5P682 2008
621.382'2--dc22   2008029481

Visit the Taylor & Francis Web site at
http://www.taylorandfrancis.com

and the CRC Press Web site at
http://www.crcpress.com

Contents

Preface

This book is written for the applied scientist and engineer who wants or needs to learn about a subject but is not an expert in the specific field. It is also written to accompany a first graduate course in signal processing. In this book, we have selected the field of random discrete signal processing. The field of random signal processing is important in diverse fields such as communications, control, radar, sonar, seismology, and bioengineering.

The aim of this book is to provide an introduction to the area of discrete random signal processing and filtering of such signals with linear and nonlinear filters. Since the signals involved are random, an introduction to random variables is also covered.

Two primary appendices are given at the end of the book to supplement the material throughout the text. Appendix A is recommended for beginners as it provides simple but important instructions to MATLAB®. Appendix B provides the fundamentals of matrix algebra. This book includes a large number of Book MATLAB functions and m-files as well as numerous MATLAB functions. These MATLAB files are useful for the reader to verify the results and create a better understanding of the presented material. This book includes many computer examples to illustrate the underlying theory and applications of discrete random signal processing. Finally, at the end of the chapters several problems are presented for a deeper understanding of the presented material. Detailed solutions, suggestions, and hints are provided for the problems in the book.

This book presents several new approaches that may be more effective in presenting the desired results. Although not robust, under the correct selection of the parameters the proposed approaches provide highly improved results. It is *highly* recommended that the reader uses the MATLAB functions to find results by inserting all the different combinations of parameters so that the reader gets an overview of how much each parameter affects the results.

Section 7.8 can be skipped without any loss of continuity.

Additional material is available from the CRC Web <http://www.crcpress.com/e_products/downloads/download.asp?cat_no=89331>.

MATLAB is a registered trademark of The MathWorks, Inc. For product information, please contact:

The MathWorks, Inc.
3 Apple Hill Drive
Natick, MA 01760-2098 USA
Tel: 508 647 7000
Fax: 508-647-7001
E-mail: info@mathworks.com
Web: www.mathworks.com

Author

Alexander D. Poularikas received his PhD from the University of Arkansas and became a professor at the University of Rhode Island. Dr. Poularikas became the chairman of the engineering department at the University of Denver and then became the chairman of the electrical and computer engineering department at the University of Alabama in Huntsville.

He has published seven books and has edited two. Dr. Poularikas has served as the editor-in-chief of the Signal Processing series (1993–1997) with Artech House and is now the editor-in-chief of the Electrical Engineering and Applied Signal Processing series as well as the Engineering and Science Primer series (1998 to present) with Taylor & Francis. He was a Fulbright scholar, is a lifelong senior member of IEEE, and is a member of Tau Beta Pi, Sigma Nu, and Sigma Pi. In 1990 and 1996, he received the Outstanding Educators Award of IEEE, Huntsville Section. He is now a professor emeritus at the University of Alabama in Huntsville.

Dr. Poularikas has authored, coauthored and edited the following books:

Electromagnetics, Marcel Dekker, 1979
Electrical Engineering: Introduction and Concepts, Matrix Publishers, 1982
Workbook, Matrix Publishers, 1982
Signals and Systems, Brooks/Cole, 1985
Elements of Signals and Systems, PWS-Kent, 1988
Signals and Systems, 2nd Edition, PWS-Kent, 1992
The Transforms and Applications Handbook, CRC Press, 1995
The Handbook for Formulas and Tables for Signal Processing, CRC Press 1998, Second Edition 2000, Third Edition 2009
Adaptive Filtering Primer with MATLAB, Taylor & Francis, 2006
Signals and Systems Primer with MATLAB, Taylor & Francis, 2007
Discrete Random Signal Processing and Filtering Primer with MATLAB, Taylor & Francis, 2009

Abbreviations

AIC	Akaike information criterion
AR	autoregressive
ARMA	autoregressive-moving average
cdf	cumulative distribution function
$comb_T(t)$	comb function, sum of infinite delta functions spaced T apart
CRLB	Cramer–Rao lower bound
DFT	discrete Fourier transform
DTFT	discrete-time Fourier transform
EDNSS	error data normalized step size
ENSS	error normalized step size
FPE	final prediction error
FT	Fourier transform
IDTFT	inverse discrete-time Fourier transform
iid	independent and identically distributed
LMS	least mean square
LS	least squares
LTI	linear time invariant
MA	moving average
MEM	maximum entropy method
MLE	maximum likelihood estimator
MMSE	minimum mean square error
MSE	mean square error
MV	minimum variance
MVU	minimum variance unbiased
MVUE	minimum variance unbiased estimator
NLMS	normalized least mean square
pdf	probability density function
PSD	power spectral density
rv	random variable
RVSS	robust variable step size
RW	random walk
SCWF	self-correcting Wiener filter

VSLMS	variable step-size LMS
WGN	white Gaussian noise
WN	white noise
WSS	wide-sense (or weakly) stationary
YW	Yule–Walker

chapter one

Fourier analysis of signals

1.1   Introduction

One of the most fundamental information of any signal is its cyclic or oscillating activity. Our main interest is to determine the sinusoidal components of the signal, the range of their amplitude and frequency values.

Fourier analysis provides us with different tools, which are appropriate for particular type of signals. If the signal is periodic and deterministic, we use Fourier series analysis. If the signal is finite and deterministic, we use the Fourier transform (FT). For discrete time signals the discrete FT (DFT) is used, whereas the random signals need special approaches for their frequency content determination. Figure 1.1 shows a deterministic periodic, a deterministic finite and a random signal.

1.2   Fourier transform (FT)

The FT is used to find the frequency content of those signals that, at least, satisfy the Dirichlet conditions, which are (1) the signal has finite values of its maximum and minimum within a finite interval, (2) the signal has finite number of discontinuities, and (3) $\int_{-\infty}^{\infty} |f(t)| dt < \infty$. The FT pair is ($\omega = 2\pi f$ has units in rad/s and f is the frequency having units cycles/s = Hz):

$$F(\omega) \triangleq \mathfrak{F}\{f(t)\} = \int_{-\infty}^{\infty} f(t) e^{-j\omega t} dt \qquad (1.1)$$

$$f(t) = \frac{1}{2\pi} \int_{-\infty}^{\infty} F(\omega) e^{j\omega t} d\omega \qquad (1.2)$$

During the integration process we treat the imaginary factor $j = \sqrt{-1}$ as a constant. Some common FT pairs of time functions are given in Appendix 1.1 at the end of this chapter.

Example 1.1

To find the FT of $f(t) = \exp(-t)u(t)$, we use Equation 1.1.
Hence,

1

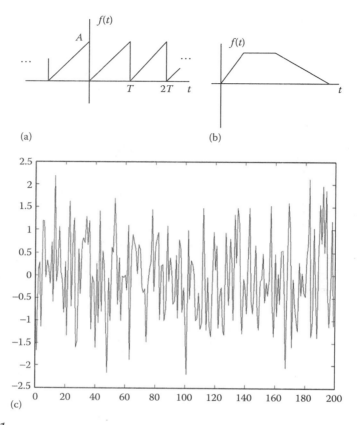

Figure 1.1

$$\mathfrak{F}\{e^{-t}u(t)\} = F(\omega) = \int_{-\infty}^{\infty} e^{-t}u(t)e^{-j\omega t}dt = \int_{0}^{\infty} e^{-(1+j\omega)t}dt = -\frac{1}{(1+j\omega)}\left[e^{-(1+j\omega)t}\right]_{0}^{\infty}$$

$$= \frac{1}{(1+j\omega)} = \frac{1}{\sqrt{(1+\omega^2)}e^{j\tan^{-1}(\omega/1)}} = \frac{1}{\sqrt{(1+\omega^2)}}e^{-j\tan^{-1}\omega}$$

$$= |F(\omega)|e^{j\varphi(\omega)} \triangleq A(\omega)e^{j\phi(\omega)}$$

where $|F(\omega)|$ is the **magnitude** spectrum and $\varphi(\omega)$ is the **phase** spectrum. Figure 1.2 shows graphically the two spectra. For this case, the frequency range is from $-\infty$ to ∞ (rad/s). The function $u(t)$ is known as the unit step function and is defined as follows:

$$u(t) = \begin{cases} 1, & t \geq 0 \\ 0, & t < 0 \end{cases} \tag{1.3}$$

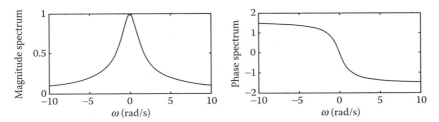

Figure 1.2

Note: *The FT operation on a deterministic continuous signal in the time domain transforms it to a continuous signal in the frequency domain. We must have in mind that only the positive frequencies are physically realizable. Negative frequencies are involved to complete mathematical operations in complex format.*

1.3   Sampling of signals

Mathematically, the sampling of signals is accomplished by multiplying the signal with the comb$_T$ (t) function. This function is made up of delta functions, equally spaced apart, with the range, $-\infty < t < \infty$. Figure 1.3a shows a continuous signal, Figure 1.3b shows the comb$_T$ function, and Figure 1.3c

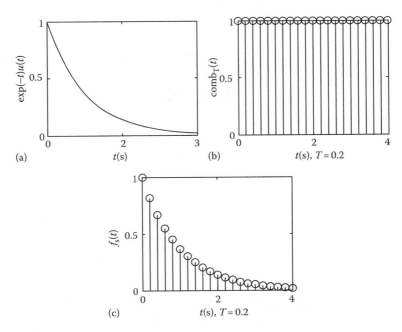

Figure 1.3

shows the resulting sampled function. In the figure, the comb function extends from $-\infty$ to ∞ and the function extends from 0 to ∞. Therefore, the sampled function is given by

$$f_s(t) = f(t)\,\text{comb}_T(t) = f(t)\sum_{n=-\infty}^{\infty}\delta(t-nT) = \sum_{n=-\infty}^{\infty}f(nT)\delta(t-nT)$$

$$= \sum_{n=0}^{\infty}f(nT)\delta(t-nT) \qquad (1.4)$$

The sampling process is based on the delta function properties, which are

$$\delta(t) = \begin{cases} 0, & t \neq 0 \\ \int\limits_{-\infty}^{\infty}f(t)\delta(t)\mathrm{d}t = f(0) & \text{or} \quad \int\limits_{-\infty}^{\infty}f(t)\delta(t-a)\mathrm{d}t = f(a) \end{cases} \qquad (1.5a)$$

$$f(t)\delta(t-a) = f(a)\delta(t-a) \qquad (1.5b)$$

We can also represent the delta function by sequences which tend to a delta function as $n \to \infty$. Table 1.1 gives such sequences.

Table 1.1   Sequences Defining the Delta Function

$\delta_n(t)$
$2n\dfrac{\sin 2\pi nt}{2\pi nt}$
$\sqrt{n}\,\mathrm{e}^{-\pi nt^2}$
$\dfrac{n}{2}\mathrm{e}^{-n|t|}$
$\dfrac{1}{\pi}\dfrac{n}{n^2t^2+1}$
$\dfrac{n\sin^2(\pi nt)}{(\pi nt)^2}$

Since we are always forced to sample continuous functions to be processed by computers, it is natural to ask the question if the spectrum of the original signal is modified or not by the sampling process. Hence, we write

$$F_s(\omega) = \mathfrak{F}\{f_s(t)\} = \mathfrak{F}\{f(t)\text{comb}_T(t)\} = \frac{1}{2\pi}F(\omega) * \frac{2\pi}{T}\text{COMB}_{\omega_s}(\omega)$$

$$= \frac{1}{T}F(\omega) * \text{COMB}_{\omega_s}(\omega) = \frac{1}{T}F(\omega) * \left[\sum_{n=-\infty}^{\infty}\delta(\omega - n\omega_s)\right], \quad \omega_s = \frac{2\pi}{T} \quad (1.6)$$

The $\text{COMB}_{\omega_s}(\omega)$ is an infinity set of delta functions being apart by the sampling frequency ω_s, and the function $(2\pi/T)\text{COMB}_{\omega_s}(\omega)$ (see Problem 1.3.1) is the FT of the $\text{comb}_T(t)$ function. The symbol "*" indicates convolution of functions. The above expansion is based on the FT property which states (see Appendix 1.2): the FT of the product of two functions is equal to the convolution of their FTs divided by 2π. The convolution of any two functions is defined as follows:

$$g(t) = f(t) * h(t) = \int_{-\infty}^{\infty} f(x)h(t-x)dx \qquad (1.7)$$

To find the convolution of two functions in the time domain, for example, we first define another domain, say x. Next, we substitute the variable t in one of the functions with the new variable x, $f(x)$. The function is identical in the x-domain as it was in the time domain. Next, in the other function wherever we see the variable t we substitute it with the variable $(t - x)$. This transformation creates a function in the x-domain which is a flipped version of that in the time domain and shifted by an amount of t. Then, we multiply these two functions and integrate. The area under the new function $f(x)h(t - x)$ is the value of the output $g(t)$ at time t. By changing the time t from $-\infty$ to ∞ we obtain the output of the convolution process, which is the function $g(t)$. Of course, the infinitesimal shifting is done automatically during integration.

Example 1.2

The convolution of the two signals $f(t) = \exp(-t)u(t)$ and $h(t) = u(t)$ is equal to

$$g(t) = \int_{-\infty}^{\infty} f(x)h(t-x)dx = \int_{-\infty}^{\infty} e^{-x}u(x)u(t-x)dx = \int_{0}^{t} e^{-x}dx = 1 - e^{-t} \qquad (1.8)$$

The functions in the t- and x-domain are shown in Figure 1.4. Note that the first function is reproduced in

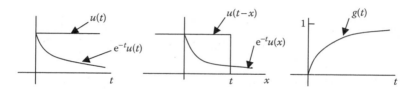

Figure 1.4

the x domain by simply substituting t with x. In the
second function, we substituted t with $t - x$. Since we
integrate with respect to x, the substitution produces a
function that is reflected along the ordinate axis and is
shifted by t. The value of t varies from $-\infty$ to ∞. The
limits of the second integral are defined by applying
the properties of the unit step function. $u(x)$ is equal to
0 for $x < 0$ and equal to 1 for $x > 0$. Since $u(t - x)$ is
flipped, it is equal to 0 for $x > t$ and equal to 1 for $x < t$.
Therefore, the integration takes place within the range
where the two functions are overlapped. For this case, we
observe two ranges of t. The first range is $-\infty < t < 0$, and
the two functions do not overlap and, hence, $g(t) = 0$. In
the other range, $0 < t < \infty$, the overlap is between 0
and t and, thus, the convolution integral is equal to
$g(t) = 1 - \exp(-t)$, $0 < t < \infty$.   ■

Let us investigate the convolution between a shifted delta function and
another continuous function $f(t)$. By definition, we write

$$f(t) * \delta(t - t_0) = \int_{-\infty}^{\infty} f(x)\delta(t - x - t_0)dx = f(t - t_0) \qquad (1.9)$$

The integration was performed using the delta function property, which
states: instead of performing integration, we first find the point where the
delta function is located on the x-axis and this value is introduced into the
rest of the functions contained in the integrand.

Note: *The convolution of a shifted delta function with any other function produces that
function exactly but shifted by the amount the delta function was initially shifted.*

Returning back to Equation 1.6, and remembering the properties of
the convolution of a function with a delta function, we observe that the
spectrum of the sampled function $F_s(\omega)$ is the sum of an infinite number of
the spectrum of the function $f(t)$, $F(\omega)$, each one shifted by the sampling frequency $\omega_s = 2\pi / T$. Figure 1.5 shows the spectrum of a function $f(t)$ and the

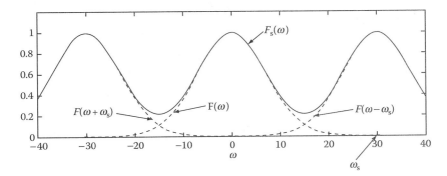

Figure 1.5

spectrum of its sampled form $f_s(t)$ using only three replicas and 30 rad/s sampling frequency. From Figure 1.5, we note that if the sampling frequency increases, equivalently when the sampling time decreases, the spectrums are spreading more and more apart so that only negligible overlapping occurs; this gives us the opportunity to recover a close approximation of the original signal. Because the spectrum of a finite-time function is infinite, it is impossible to recover the total signals spectrum and, thus, always we will recover a distorted one. The sampling theorem defines the conditions under which the signal can be recovered without distortion. The theorem states:

Sampling theorem: *For any band-limited signal, with its highest frequency ω_N (Nyquist frequency), can be recovered completely from its sampled function if the sampling frequency ω_s is at least twice the Nyquist frequency ω_N, or the sampling time must be at least half the Nyquist sampling time $2\pi/\omega_N$.*

1.4   Discrete-time Fourier transform (DTFT)

At the limit as $T \to 0$, we can approximate the Fourier integral as follows (see Problem 1.4.1):

$$F(\omega) = T \sum_{n=-\infty}^{\infty} f(nT)e^{-j\omega nT}, \quad \omega = \text{rad/s} \tag{1.10}$$

Since $\exp(-j(\omega + (2\pi/T))nT) = \exp(-j\omega nT)\exp(-j2\pi n) = \exp(-j\omega nT)$, this result indicates that the spectrum of the sampled function with sampling time T is periodic with period $2\pi/T$. Therefore, the inverse DTFT is given by

$$f(nT) = \frac{1}{2\pi} \int_{-\pi/T}^{\omega/T} F(\omega)e^{j\omega nT} d\omega \tag{1.11}$$

and, hence, Equations 1.10 and 1.11 constitute the DTFT pair.

Example 1.3

Find and plot the magnitude of the DTFT spectrum for the sequence $f(n) = 0.9^n u(n)$.

Solution

Using Equation 1.10 with $T = 1$, we obtain the spectrum

$$F(\omega) = \sum_{n=0}^{\infty} 0.9^n e^{-jn\omega} = \sum_{n=0}^{\infty} (0.9 e^{-j\omega})^n = \frac{1}{1 - 0.9 e^{-j\omega}}$$

$$= \frac{1}{1 - 0.9(\cos(\omega) - j\,\sin(\omega))}$$

$$= \frac{1}{\sqrt{(1 - 0.9\cos(\omega))^2 + (0.9\sin(\omega))^2}} e^{-j\tan^{-1}\frac{0.9\sin(\omega)}{1 - 0.9\cos(\omega)}} \quad (1.12)$$

$$= A(\omega) e^{j\phi(\omega)}$$

which is given in the polar form, amplitude and phase spectra. The expansion of the summation indicates that it is a geometric infinite sequence of the form $(1 + x + x^2 + x^3 + \cdots) = 1/(1 - x)$. The sampling frequency in this case is $2\pi/1$. The magnitude of the spectrum is shown in Figure 1.6. Note the periodicity of the spectrum. Mathematics inform us that to have a convergence sequence and be able to sum the above infinite series the absolute value of $0.9\exp(-j\omega)$ must be less than 1, or $|0.9\exp(-j\omega)| = |0.9||\exp(-j\omega)| = 0.9 < 1$, which verifies the requirement. ■

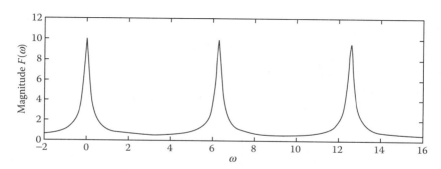

Figure 1.6

Note: *The signal is discrete in the time domain and its DTFT is a continuous function of ω in the frequency domain and is periodic.*

1.5   Discrete Fourier transform (DFT)

With the development of an efficient computational procedure, known as the **fast Fourier transform** (FFT), the DFT is used extensively. As a practical matter, we are only able to manipulate a certain length of a discretized continuous signal. That is, suppose that the data sequence is available only within a finite time window from $n = 0$ to $N - 1$. The transform of the discrete signal is also discrete having the same N values. The samples in the frequency domain are located at intervals of $2\pi/NT$ apart, where T (sampling time) is the time interval between sample points in the time domain. Hence, we define the DFT of a sequence of N samples, $\{f(nT)\}$ for $0 \le n \le N - 1$, by the relation:

$$\begin{aligned}
F(k\Omega_b) \triangleq \mathfrak{F}_D\{f(nT)\} &= T\sum_{n=0}^{N-1} f(nT)e^{-j2\pi knT/NT} \\
&= T\sum_{n=0}^{N-1} f(nT)e^{-j\Omega_b Tkn}, \quad k = 0,1,\dots,N-1
\end{aligned} \tag{1.13}$$

where
   N is the number of sample values
   T is the sampling time interval or sampling time (time bin)
   $(N - 1)T$ is the signal length
   $\Omega_b = (\omega_s/N) = (2\pi/NT)$ is the frequency sampling interval (frequency bin)
   ω_s is the sampling frequency
   $e^{-j\Omega_b T}$ is the Nth principal root of unity

Note: *With this specification of the DFT and Ω_b, there are only N distinct values computable by Equation 1.13.*

The inverse DFT (IDFT) is given by the relation:

$$\begin{aligned}
f(nT) = \mathfrak{F}_D^{-1}\{F(k\Omega_b)\} &= \frac{1}{NT}\sum_{k=0}^{N-1} F(k\Omega_b)e^{j2\pi nkT/NT} \\
&= \frac{1}{NT}\sum_{k=0}^{N-1} F(k\Omega_b)e^{jn\Omega_b kT}, \quad n = 0,1,2,\dots,N-1
\end{aligned} \tag{1.14}$$

It turns out that both sequences are **periodic** with period N. This stems from the fact that the exponential function is periodic as shown below

$$e^{\pm j2\pi nk/N} = e^{\pm j2\pi k(n+N)/N}, \quad e^{\pm j2\pi nk/N} = e^{\pm j2\pi n(k+N)/N}, \quad k,n = 0,1,2,\dots,N-1 \tag{1.15}$$

1.5.1   Properties of DFT

To simplify the proofs, without any loss of generality, we set $T = 1$ and $k2\pi/N = k$.

Linearity

The DFT of the function $f(n) = ax(n) + by(n)$ is

$$F(k) = \mathfrak{F}_D\{ax(n) + by(n)\} = a\mathfrak{F}_D\{x(n)\} + b\mathfrak{F}_D\{y(n)\} = aX(k) + bY(k) \quad (1.16)$$

This property is the direct result of Equation 1.13.

Symmetry

If $f(n)$ and $F(k)$ are a DFT pair, then (see Problem 1.5.1):

$$\mathfrak{F}_D\left\{\frac{1}{N}F(n)\right\} = f(-k) \quad (1.17)$$

Time shifting

For any real integer m (see Problem 1.5.2):

$$\mathfrak{F}_D\{f(n-m)\} = F(k)e^{-j2\pi mk/N} \quad (1.18)$$

The above equation indicates that, if the function is shifted in the time domain by m units of time to the right, its spectrum magnitude does not change. However, its phase spectrum is changing by the addition of a linear phase factor equal to $-2\pi m/N$.

Frequency shifting

For any integer m (see Problem 1.5.3):

$$f(n)e^{j2\pi nm} = \mathfrak{F}_D^{-1}\{F(k-m)\} \quad (1.19)$$

Time convolution

The discrete convolution is defined by the expression:

$$y(nT) = f(nT) * h(nT) = T\sum_{m=-\infty}^{\infty} f(mT)h(nT - mT)$$

$$y(n) = f(n) * h(n) = \sum_{m=-\infty}^{\infty} f(m)h(n - m) \quad (1.20)$$

where both discrete signals are periodic with the same period N

$$f(n) = f(n + pN), \quad p = 0, \pm 1, \pm 2, \ldots, \qquad h(n) = h(n + pN), \quad p = 0, \pm 1, \pm 2, \ldots$$
(1.21)

This type of convolution is known as **circular** or **cyclic** convolution. The DFT of the convolution expression yields (see Problem 1.5.4):

$$Y(k) = \mathfrak{F}_D\{f(n) * h(n)\} = F(k)H(k)$$
(1.22)

Figure 1.7 shows graphically the convolution of the following two periodic sequences, $f(n) = \{1, 2, 3\}$ and $h = \{1, 1, 0\}$. Figure 1.7a shows the two sequences at 0 shifting and the result is given by the sum of the products: $1 \times 1 + 2 \times 0 + 3 \times 1 = 4$. Next, we shift the inside circle by one step counterclockwise and the output is given by the sum of the following products: $1 \times 1 + 2 \times 1 + 3 \times 0 = 3$ and so on. Note that $y(n)$ is periodic and, since $F(k)$ and $H(k)$ are periodic, $Y(k)$ is also periodic.

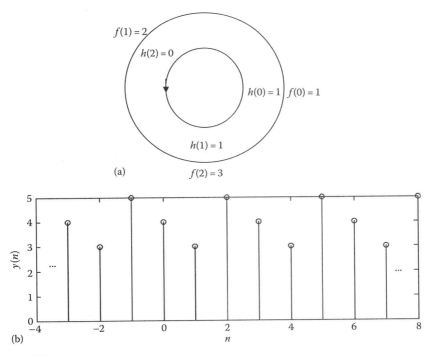

(a)

(b)

Figure 1.7

Another approach to the evaluation of circular convolution is based on Equation 1.20. This equation can be written in the form given below for the particular case when each sequence has three terms.

$$\begin{aligned}
y(0) &= f(0)h(0) + f(1)h(2) + f(2)h(1) \\
y(1) &= f(0)h(1) + f(1)h(0) + f(2)h(2) \\
y(2) &= f(0)h(2) + f(1)h(1) + f(2)h(0)
\end{aligned} \tag{1.23}$$

This set of equation can be written in matrix form as follows:

$$[y]^{\mathrm{T}} = \begin{bmatrix} y(0) & y(1) & y(2) \end{bmatrix} = \begin{bmatrix} f(0) & f(1) & f(2) \end{bmatrix} \begin{bmatrix} h(0) & h(1) & h(2) \\ h(2) & h(0) & h(1) \\ h(1) & h(2) & h(0) \end{bmatrix} \tag{1.24}$$

If we want to produce a linear convolution, using the matrix format, we must pad with zeros of the sequences so that their total length must be equal to $N = F + H - 1$, where F and H are the number of elements of the corresponding sequences.

Example 1.4

Consider the two periodic sequences $f(n) = \{1, -1, 4\}$ and $h(n) = \{0, 1, 3\}$. Verify the convolution property for $n = 2$.

Solution

In this case $T = 1$, $\Omega_b = 2\pi/3$, and

$$\begin{aligned}
y(2) &= \sum_{m=0}^{N-1} f(m)h(n-m) = \mathfrak{F}_D^{-1}\{F(k)H(k)\} \\
&= \frac{1}{N} \sum_{m=0}^{N-1} F(k)H(k)e^{j2\pi k2/N}
\end{aligned}$$

First, we find the summation:

$$\begin{aligned}
\sum_{m=0}^{2} f(m)h(2-m) &= f(0)h(2) + f(1)h(1) + f(2)h(0) \\
&= 1 \times 3 + (-1) \times 1 + 4 \times 0 = 2
\end{aligned}$$

Next, we obtain $F(k)$ and $H(k)$:

$$F(0) = \sum_{n=0}^{2} f(n)e^{-j2\pi 0n/3} = f(0) + f(1) + f(2) = 1 - 1 + 4 = 4$$

$$F(1) = \sum_{n=0}^{2} f(n)e^{-j2\pi 1 n/3} = f(0) + f(1)e^{-j2\pi/3} + f(2)e^{-j2\pi 4/3}$$

$$= -0.5 + j5 \times 0.866$$

$$F(2) = \sum_{n=0}^{2} f(n)e^{-j2\pi 2n/3} = f(0) + f(1)e^{-j2\pi 2/3} + f(2)e^{-j2\pi 4/3}$$

$$= -0.5 - j5 \times 0.866$$

Similarly, we obtain

$$H(0) = 4, \qquad H(1) = -2 + j2 \times 0.866,$$
$$H(2) = -2 - j2 \times 0.866$$

The second summation given above becomes

$$\frac{1}{3}\sum_{k=0}^{2} F(k)H(k)e^{j2\pi 2k/3} = \frac{1}{3}\left[16 + 11.55e^{j115.7} + 11.55e^{j604.3}\right] = 2$$

This result shows the validity of the DFT of the convolution. ∎

Frequency convolution

The frequency convolution property is given by (see Problem 1.5.5):

$$\mathfrak{F}_D\{g(n)\} = \mathfrak{F}_D\{f(n)h(n)\} = \frac{1}{N}\sum_{m=0}^{N-1} F(m)H(k-m) \qquad (1.25)$$

where $g(n) = f(n)h(n)$.

Parseval's theorem

$$\sum_{n=0}^{N-1} f^2(n) = \frac{1}{N}\sum_{k=0}^{N-1} |F(k)|^2 \qquad (1.26)$$

1.5.2   *Effect of sampling time* T

Let us deduce the DFT of the continuous function:

$$f(t) = e^{-t}, \quad 0 \le t \le 1$$

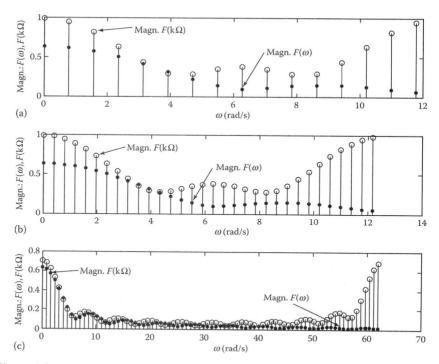

Figure 1.8

for the following three cases:

   a. $T = 0.5, NT = 8$
   b. $T = 0.5, NT = 16$
   c. $T = 0.1, NT = 8$

Case a: In this case, we find $N = 8/T = 8/0.5 = 16$ and $\Omega_b = 2\pi/(NT) = 2\pi/8$. The fold-over frequency for this case is at $N/2 = 8$, or at $8 \times \Omega_b = 8 \times [2\pi/(0.5 \times 16)] = 2\pi$ rad/s. The Fourier amplitude spectrums for the continuous and discrete case are shown in Figure 1.8a. The exact spectrum is found to be equal to: $F(\omega) = (1 - \exp(1 - (1 + j\omega)))/(1 + j\omega)$. Although the spectrum of the continuous signal is continuous and infinite in length, we only present the values which correspond to the spectrum of the discrete case. We observe that the errors between these two spectrums are considerable up to the fold-over frequency, and these errors become larger for the rest of the range. The MATLAB® program given below was used to find the exact and DFT magnitude values of the spectrum and were plotted in Figure 1.8. Note that the number of function values used in the calculations were three, $\{f(0), f(0.5), f(1)\}$. However, since $N = 16$ the MATLAB function **fft(·)** added 13 zeros, thus creating a vector with 16 values. The m-file below produces Figure 1.8. Similarly, we can graph cases (b) and (c).

Book MATLAB m-file: eff_sampl_time

```
%Book m-file:eff_sampl_time
w1=0:pi/4:4*pi-(pi/4);
fw1=abs((1-exp(-(1+j*w1)))./(1+j*w1));%abs(.) is a
              %MATLAB function;
t1=0:0.5:1;
ft1=exp(-t1);
dft1=0.5*abs(fft(ft1,16));%fft(.) is the fast Fourier
              %transform MATLAB function;
subplot(3,1,1);stem(w1,dft1,'k');% 'k' instructs MATLAB
              %to produce a black graph;
hold on; plot(w1,fw1,'.k', 'markersize', 15);% '.k'
              %produces black dots of magnitude 15;
xlabel('\omega rad/s');ylabel('Magn: F(\omega),...
  F(\Omega_b)');
```

Calling the m-files: eff_sampl_time_1 and eff_sampl_time_2 will produce Figure 1.8b and c corresponding to case (b) and (c) discussed below.

Case b: In this case, we do not change the sampling time, $T = 0.5$, but we double the number $NT = 16$. From these data, we obtain $N = 32$. Hence, the fold-over frequency is $16 \times 2\pi/16 = 2\pi$, which is identical to that found in case (a) above. Therefore, the discrete spectrum is periodic with a period of 2π. We observe in Figure 1.8b that although we doubled NT, doubling the number of points in the frequency domain for the same range of frequencies, the accuracy of the discrete spectrum has not improved and, in the same time, the largest frequency which approximates the exact spectrum is the same, $\pi/0.5 = 2\pi$. The figure can be plotted easily by a slight change of the program given above.

Case c: In this case, $NT = 8$, but the sampling time was decreased to 0.1 value. This decrease of sampling time resulted in the increase of function sampling points from 3 to 11. From Figure 1.8c, we observe that the accuracy has increased and the approximation to the maximum frequency has also increased to $\pi/0.1 = 10\pi$. The frequency bins are equal to $2\pi/8 = \pi/4$ and the fold-over frequency is equal to $(\pi/4)40 = 10\pi$. The total number of points were $N = 8/0.1 = 80$.

Note: *Based on the above results, we conclude (1) keeping the sampling time T constant and increasing the number of values of the sampled function by padding it with zeros increases the number of frequency bins and, thus, making the spectrum better defined but its* **accuracy does not increase.** *(2)* **Decreasing the sampling time T results in better accuracy of the spectrum.** *It further extends the range in simulating the exact spectrum using the DFT approach and this range is π/T.*

1.5.3   Effect of truncation

Because the DFT uses finite number of samples, we must be concerned about the effect that the truncation has on the Fourier spectrum, even if the original function extends to infinity. Specifically, if the signal $f(t)$ extends beyond the total sampling period NT, the resulting frequency spectrum is an approximation to the exact one. If, for example, we take the DFT of a truncated sinusoidal signal, we find that the FT consists of additional lines that are the result of the truncation process. Therefore, if N is small and the sampling covers neither a large number nor an integral number of cycles of the signal, a large error in spectral representation can occur. This phenomenon is known as **leakage** and is the direct result of truncation. Since the truncated portion of the signal is equal to $f(t)p_a(t)$, the distortion is due to the presence of the rectangular window. This becomes obvious when we apply the FT property: $\mathfrak{F}_D\{f(nT)p_a(nT)\} = F(k\Omega_b) * P_a(k\Omega_b)$. This expression indicates that the exact spectrum is modified due to the convolution operation. It turns out that as the width of the pulse becomes very large, its spectrum resembles a delta function and, hence, the output of the convolution is close to the exact spectrum. Figure 1.9a and b presents pictorially the effect of truncation. In Figure 1.9a (curve c), the resulting spectrum shows the largest undulations because the rectangular window is the shortest. As the window becomes wider, its effect during the convolution becomes less important as shown in Figure 1.9a (curve b) curve a is due to exponential function but with a window quite larger than 1.2s: Figure 1.9b is the spectrum of the rectangular window for length 1.2s. The spectrums were plotted in continuous format for better visualization. Based on the above analysis, a question arises: are any other types of windows, which may distort less the signal spectrum? Table 1.2 includes some other proposed windows.

1.5.4   Windowing

The reader must run a few examples using different types of windows. Meanwhile, the reader should also find the spectrums of the windows and get familiar with level of their side-lobes and, in particular, the level of the

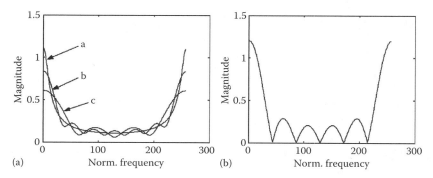

Figure 1.9

Table 1.2 Windows in the Time Domain

Data Window

Rectangular

$w(n) = 1, \quad 0 \le n \le N - 1$

Triangular–Bartlett

$$w(n) = \begin{cases} 2n/(N-1), & 0 \le n \le ((N-1)/2) \\ 2-(2n/(N-1)), & ((N-1)/2) \le n \le N-1 \end{cases}$$

Blackman

$$w(n) = 0.42 - 0.5\cos\left(\frac{2\pi n}{N-1}\right) + 0.08\cos\left(\frac{4\pi n}{N-1}\right), \quad 0 \le n \le N-1$$

Hamming

$$w(n) = 0.54 - 0.46\cos\left(\frac{2\pi n}{N-1}\right), \quad 0 \le n \le N-1$$

Hanning–Tukey

$$w(n) = \frac{1}{2}\left(1 - \cos\left(\frac{2\pi n}{N-1}\right)\right), \quad 0 \le n \le N-1$$

$\cos^\alpha(\cdot)$ window

$$w(n) = \sin^\alpha\left(\frac{n}{N}\pi\right), \quad 1 \le \alpha \le 4$$

Kaiser–Bessel window

$$w(n) = \frac{I_0\left[\pi\alpha\sqrt{1.0 - (n/(N/2))^2}\right]}{I_0[\pi\alpha]}, \quad 0 \le |n| \le N/2$$

$I_0[x] = \sum_{k=0}^{\infty}\left[\frac{(x/2)^k}{k!}\right]^2$ = zero-order modified Bessel function and α ranges between 1 and 4.

The expansion of $I_0[x]$ to about 25 terms is sufficient.

first side-lobe. The plots should be done in a log-linear scale. Furthermore, the width of the main lobe plays an important role during the convolution process in the frequency domain. The thinner the main lobe is the better approximation we obtain of the desired spectrum. Both, the main lobe and the level of the first side-lobe play an important role in the attempt to produce the smallest distortion of the spectrum of the truncated signal.

1.6   Resolution

One interesting question is about the ability to resolve two frequencies which are close together. The main lobe of the window plays an important effect in the resolution of frequencies. Figure 1.10 shows the effect of the window that affects the resolution. The function we used is $f(n) = \cos(0.2\pi n) + \cos(0.4\pi n) + \cos(0.45\pi n)$. The resolution is understood by comparing the values $\Delta\omega$ with the unit bin in the frequency domain, which is equal to $2\pi/NT$. In this case, we used $T = 1$. In Figure 1.10a, we find that $(0.45 - 0.4)\pi \ll 2\pi/16$ and it is obvious that the two closed spaced frequencies cannot be resolved. In Figure 1.10b, we find that 0.05π is equal to about $2\pi/32$, and it is apparent that a separation of the two frequencies starting to be resolved. However, in the third case where $0.05\pi \gg 2\pi/128$, the resolution is perfect. The Book MATLAB m-file below plots Figure 1.10.

Book MATLAB m-file: non_rand_resolu

```
%resolution m-file: non_rand_resolu.m
n1=0:15;
f1=cos(0.2*pi*n1)+cos(0.4*pi*n1)+cos(0.45*pi*n1);
n2=0:31;
f2=cos(0.2*pi*n2)+cos(0.4*pi*n2)+cos(0.45*pi*n2);
n3=0:127;
f3=cos(0.2*pi*n3)+cos(0.4*pi*n3)+cos(0.45*pi*n3);
ftf1=fft(f1,128);ftf2=fft(f2,128);ftf3=fft(f3,128);
w=0:2*pi/128:2*pi-(2*pi/128);
subplot(2,3,1);plot(w,abs(ftf1),'k');xlabel('\omega');
ylabel('Magn. F(k\Omega_s)');title('N=16, a)');
subplot(2,3,2);plot(w,abs(ftf2),'k');xlabel('\omega');
ylabel('Magn. F(k\Omega_s)');title('N=32, b)');
subplot(2,3,3);plot(w,abs(ftf3),'k');xlabel('\omega');
ylabel('Magn. F(k\Omega_s)');title('N=128, a)');
```

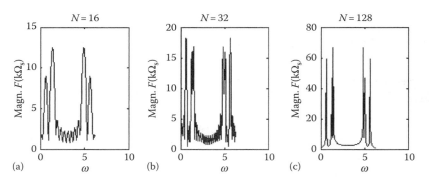

Figure 1.10

1.7   Continuous linear systems

1.7.1   Properties of linear time-invariant (LTI) systems

Every **physical system** is broadly characterized by its ability to accept an **input** (voltage, current, pressure, magnetic fields, electric fields, etc.) and to produce an **output response** to this input. To study the behavior of a system, the procedure is to model mathematically each element that the system comprises and then consider the interconnected array of elements. The analysis of most systems can be reduced to the study of the relationship among certain inputs and resulting outputs. The interconnected system is described mathematically, the form of the description being dictated by the domain of description, time or frequency domain.

In our studies, we shall deal with LTI systems whose input may be deterministic or random. We will be interested to investigate systems whose initial conditions are, in general, zero. For electrical systems, we will assume that they are not currents through the inductors or charges across the capacitors.

A simple *RL* circuit with its block representation is shown in Figure 1.11. Applying the Kirchhoff's voltage law, we obtain

$$L\frac{di(t)}{dt} + Ri(t) = v(t) \quad \text{or} \quad \frac{di(t)}{dt} + \frac{R}{L}i(t) = \frac{1}{L}v(t), \quad t \geq 0 \tag{1.27}$$

To proceed, we multiply both sides of this equation by exp(tR/L), the form is suggested from the homogeneous solution of the above equation exp($-tR/L$). We recognize that we can write the result as follows:

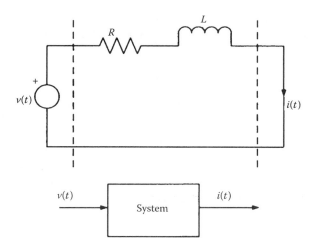

Figure 1.11

$$\frac{d}{dt}[\exp(tR/L)i(t)] = \exp(tR/L)\frac{v(t)}{L} \tag{1.28}$$

Integrate both sides of the above equation from 0 to t to obtain

$$e^{(R/L)t}i(t)\Big|_0^t = \frac{1}{L}\int_0^t e^{(R/L)x}i(x)dx \quad \text{or} \quad e^{(R/L)t}i(t) - i(0) = \frac{1}{L}\int_0^t e^{(R/L)x}i(x)dx$$

This result is written in the form:

$$i(t) = \underbrace{i(0)e^{-(R/L)t}}_{\text{Zero input solution}} + \underbrace{\frac{1}{L}\int_0^t i(x)e^{-(R/L)(t-x)}\,dx}_{\text{Zero state solution}} \tag{1.29}$$

Since we shall be dealing with situations where the zero initial conditions are zero, implies that the zero state solution will be zero. The desired solution will be the zero state one. From Equation 1.29, we observe that the zero state solution is of the general form:

$$y(t) \triangleq x(t) * h(t) = \int_{-\infty}^{\infty} x(\tau)h(t-\tau)d\tau = \int_{-\infty}^{\infty} x(t-\tau)h(\tau)d\tau \equiv \text{Convolution} \tag{1.30}$$

known as the **convolution** between the functions $x(t)$ and $h(t)$. To obtain the convolution of two functions, we go to a new domain x. To one of the functions, we substitute t with x and flip the other function in the new domain and shift it by t (just substitute t with $t - x$). Next, we multiply the two functions and find (integral) the area. This result is the output $y(t)$ at the time shift t. We repeat this procedure for all ts and, hence, we find the output.

Note: *From Equation 1.29, we conclude that the output of LTI system, with zero initial conditions, is equal to the convolution of its input and its impulse response. The impulse response of a system is its output if its input is the delta function. Hence, we write*

$$\boxed{\begin{array}{l} y(t) \triangleq \text{output} = \int_{-\infty}^{\infty} f(x)h(t-x)dx, \quad f(t) = \text{input}, \\[2mm] \phantom{y(t) \triangleq \text{output} = } h(t) = \text{impulse response} \end{array}} \tag{1.31}$$

Example 1.5

The impulse response of an RL circuit in series is found by introducing a delta function as its input. Hence, Equation 1.27 takes the form:

$$\frac{dh(t)}{dt} + \frac{R}{L}h(t) = \delta(t) \tag{1.32}$$

where $h(t)$ is the impulse response of the system and has the units of current (Amperes). Next, if we take the FT of both sides of the above equation, we find the following result:

$$H(\omega) = \frac{1}{(R/L) + j\omega} \tag{1.33}$$

From the FT tables, we find the inverse transform to be

$$h(t) = e^{-(R/L)t}, \quad t \geq 0 \tag{1.34}$$

■

Example 1.6

The convolution of the functions $x(t) = u(t)$ and $h(t) = e^{-t}u(t)$ is found by considering the different ranges of t. Let us shift and flip the $x(t)$ function. The convolution integral takes the form:

$$y(t) = \int_{-\infty}^{\infty} u(t-x)e^{-x}u(x)\mathrm{d}x \tag{1.35}$$

For the range of $t < 0$, the two functions do not overlap and, hence, the integrand is zero and, therefore, the output is zero for this range. For the range of $t \geq 0$, the two functions in the x domain are those shown in Figure 1.12, and the integral becomes

$$y(t) = \int_{0}^{t} e^{-x}\mathrm{d}x = (1 - e^{-t}), \quad t \geq 0 \tag{1.36}$$

■

Based on the convolution property of the LTI functions and the FT property of the convolution of two functions, we obtain the frequency spectrum of the output to be equal to the multiplication of the input spectrum and the impulse response spectrum. Hence, Equation 1.37 becomes

$$\boxed{Y(\omega) = F(\omega)H(\omega)} \tag{1.38}$$

1.8   Discrete systems—linear difference equations

Discrete systems can be given in two ways: in one way, the system is represented directly by a difference equation (discrete form) and, the other way,

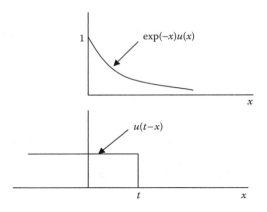

Figure 1.12

the system is given in an analog form and it is asked to be transformed in its equivalent discrete form.

To be able to transform an analog, first- and second-order systems, we need the following approximations:

$$\frac{\mathrm{d}x(t)}{\mathrm{d}t} \cong \frac{x(nT)-x(nT-T)}{T}, \quad \frac{\mathrm{d}^2x(t)}{\mathrm{d}t^2} \cong \frac{x(nT)-2x(nT-T)+x(nT-2T)}{T^2} \quad (1.39)$$

If we are dealing with an integrodifferential equation, we approximate the integral as follows:

$$y(nT) = \int_0^{nT} x(t)\mathrm{d}t = \int_0^{nT-T} x(t)\mathrm{d}t + \int_{nT-T}^{T} x(t)\mathrm{d}t \cong y(nT-T)+Tx(nT) \quad (1.40)$$

assuming that the end of integration is an exact multiple of T. Therefore, the *RL* analog system, described by Equation 1.27, is transformed to an equivalent discrete system as follows:

$$i(nT) - \frac{1}{1+(R/L)T}i(nT-T) = \frac{T}{L}\frac{1}{1+(R/L)T}v(nT) \quad (1.41)$$

The above equation is of the general form:

$$y(nT) + a_1 y(nT-T) + \cdots + a_p y(nT-pT) = b_0 x(nT) \quad (1.42)$$

This difference equation describes an **infinite impulse response** (IIR) discrete system. Any system described by the difference equation:

$$y(nT) = b_0 x(nT) + b_1 x(nT - T) + \cdots + b_q x(nT - qT) \tag{1.43}$$

is known as a **finite impulse response** (FIR) discrete system. A combined system is described as follows:

$$y(nT) + a_1 y(nT - T) + \cdots + a_p y(nT - pT)$$
$$= b_0 x(nT) + b_1 x(nT - T) + \cdots + b_q x(nT - qT) \tag{1.44}$$

We introduced the minus sign above so that we follow the general engineering practice.

Taking the DTFT of Equation 1.44, setting $T = 1$ and rearranging, we obtain the transfer function of the discrete system to be equal to

$$H(e^{j\omega}) \triangleq \frac{Y(e^{j\omega})}{X(e^{j\omega})} = \frac{b_0 + b_1 e^{-j\omega} + \cdots + b_q e^{-j\omega q}}{1 + a_1 e^{-j\omega} + \cdots + a_p e^{-j\omega p}} \tag{1.45}$$

If we substitute $z = e^{j\omega}$ in the above equation, we obtain the z-transform representation of the transfer function:

$$H(z) \triangleq \frac{Y(z)}{X(z)} = \frac{b_0 + b_1 z^{-1} + \cdots + b_q z^{-q}}{1 + a_1 z^{-1} + \cdots + a_p z^{-p}} = \sum_{n=0}^{\infty} h(n) z^{-n} \tag{1.46}$$

The sequence $\{h(n)\}$ is the impulse response of the discrete system and is given by

$$h(n) = \frac{1}{2\pi} \int_{-\pi}^{\pi} H(e^{j\omega}) e^{jn\omega} \, d\omega \tag{1.47}$$

From an **operational** point of view, z in Equation 1.46 may be regarded as a **shift operator** with the property

$$z^{-k} x(n) = x(n - k) \tag{1.48}$$

Hence, the difference equation (Equation 1.44) may be written in the form:

$$(1 + a_1 z^{-1} + \cdots + a_p z^{-p}) y(n) = (b_0 + b_1 z^{-1} + \cdots + b_q z^{-q}) x(n) \tag{1.49}$$

that is

$$y(n) = \frac{(b_0 + b_1 z^{-1} + \cdots + b_q z^{-q})}{(1 + a_1 z^{-1} + \cdots + a_p z^{-p})} x(n) = H(z) x(n) \tag{1.50}$$

where $H(z)$ is the **transfer function** of the discrete system. Expanding $H(z)$ in powers of z^{-1} gives

$$y(n) = \sum_{m=0}^{\infty} h(m)z^{-m}x(n) = \sum_{m=0}^{\infty} h(m)x(n-m) \qquad (1.51)$$

which is the **zero state solution** (no initial conditions present) of the difference equation.

Stability

Taking the absolute value of both sides of Equation 1.51, we obtain the relation:

$$|y(n)| = \left| \sum_{m=0}^{\infty} h(m)x(n-m) \right| \le \sum_{m=0}^{\infty} |h(m)| \, |x(n-m)| < M \sum_{m=0}^{\infty} |h(m)| < \infty \qquad (1.52)$$

The above relation indicates that for a bounded input (M, maximum value of the input), we will have a bounded output if the sum of the absolute value of the impulse response is finite.

1.9 Detrending

Sometimes the data may have a **trend component** $m(n)$, a **seasonal component** $s(n)$, and **a random component** $v(n)$. Such a signal is shown in Figure 1.13. Guessing that the trend is a fourth-order polynomial type, we set

$$x(n) = a_1 + a_2 n + a_3 n^2 + a_4 n^3 + a_5 n^4 \qquad (1.53)$$

Our task is to specify the unknowns a_1, a_2, a_3, a_4, and a_5 and, then, subtract $x(n)$ from the data so that only the random part of the data remains. To obtain the unknowns a_is, we proceed as follows:

$$
\begin{aligned}
a_1 + a_2 0 + a_3 0^2 + a_4 0^3 + a_5 0^4 &= s(0) \\
a_1 + a_1 1 + a_3 1^2 + a_4 1^3 + a_5 1^4 &= s(1) \\
a_1 + a_2 2 + a_3 2^2 + a_4 2^3 + a_5 2^4 &= s(2) \\
&\vdots \\
a_1 + a_2 N + a_3 N^2 + a_4 N^3 + a_5 N^4 &= s(N)
\end{aligned}
\qquad (1.54)
$$

The above equations can be written in the following matrix form:

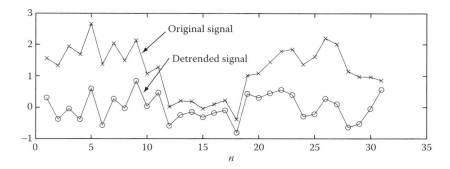

Figure 1.13

$$\begin{bmatrix} 1 & 0 & 0 & 0 & 0 \\ 1 & 1 & 1 & 1 & 1 \\ 1 & 2 & 4 & 8 & 16 \\ \vdots & \vdots & \vdots & \vdots & \vdots \\ 1 & N & N^2 & N^3 & N^4 \end{bmatrix} \begin{bmatrix} a_1 \\ a_2 \\ a_4 \\ a_5 \end{bmatrix} = \begin{bmatrix} s(0) \\ s(1) \\ s(2) \\ \vdots \\ s(N) \end{bmatrix} \quad \text{or} \quad \mathbf{Ha} = \mathbf{s} \quad \text{or} \quad \mathbf{H'Ha} = \mathbf{H's}$$

$$\text{or} \quad \boxed{\mathbf{a} = (\mathbf{H'H})^{-1}\mathbf{H's}} \qquad (1.55)$$

where **H'** means the transpose of **H**.

Next, using the vector **a** in the relation **Ha**, we obtain a closed approximation to the shift and seasonal variation of the signal. When we subtract this signal from the original one, we obtain only the random variations of the signal. The solution $(\mathbf{H'H})^{-1}\mathbf{H's}$ for the vector **a**, is known as the **least square** (LS) solution.

To obtain Figure 1.13, we used the following Book MATLAB m-file:

Book MATLAB m-File: det_and_ls

```
%Book m-file:det_and_ls
n=0:30;
s=1+sin(0.1*pi*n)+1.5*(rand(1,31)-0.5);%1x31 vector;
for m=0:30
    H(m+1,:)=[1 m m^2 m^3 m^4];%H=31x5 matrix;
end;
a=(inv(H'*H))*H'*s';
sv=a(1)+a(2)*n+a(3)*n.^2+a(4)*n.^3+a(5)*n.^4;
%sv=seasonal variation;
v=s-sv;%v=random variation of the original signal;
subplot(2,1,1);plot(s,'kx-');hold on;plot(v,'ko-');
xlabel('n');legend('Original signal','De-trended ...
   signal');
```

Problems

1.3.1 Find the FT of the comb function:

$$\text{comb}_T(t) = \sum_{n=-\infty}^{\infty} \delta(t - nT).$$

1.4.1 Verify Equation 1.10.
1.5.1 Verify Equation 1.17.
1.5.2 Verify Equation 1.18.
1.5.3 Verify Equation 1.19.
1.5.4 Verify Equation 1.22.
1.5.5 Find the DFT of $f(t) = \exp(-t)u(t)$ using sampling time $T = 0.3$ and the following three different lengths of the function: $0 \le t \le 0.6$; $0 \le t \le 1.2$; and $0 \le t \le 20$. Observe the magnitude of the function at zero frequency. Repeat the above with $T = 0.05$ and observe the difference.
1.5.6 Sample and truncate the signal $f(t) = \sin 2\pi t$, and find its spectrum for several lengths of the signal less than 1/2 of its period. Next, sample the signal for a length equal to 1/2 of its period and then, using the same sampling time, sample the signal up to 0.8 of 1/2 of its period and observe any differences.
1.5.7 Find the FFT of the signals given below and state why there is a difference in their magnitude spectra. The functions are $x1 = [1\,1\,1\,1\,1\,1\,1\,1\,0\,0\,0]$; $x2 = [1\,1\,1\,1\,1\,1\,1\,1\,1\,1\,1]$. Do not pad them with zeros.
1.7.1 Find the convolution of the functions: $f(t) = e^{-t}u(t)$; $h(t) = e^{-2t}u(t)$.

Solutions, suggestions, and hints

1.3.1 The comb function is periodic and can be expanded in the complex Fourier series form:

$$\text{comb}_T(t) = \sum_{n=-\infty}^{\infty} a_n e^{jn\omega_s t}, \quad \omega_s = \frac{2\pi}{T}, \quad a_n = \frac{1}{T} \int_{T/2}^{T/2} \delta(t) e^{-jn\omega_s t}\, dt = \frac{1}{T}$$

hence

$$\text{comb}_T(t) = \frac{1}{T} \sum_{n=-\infty}^{\infty} e^{jn\omega_s t}$$

$$\mathfrak{F}\{\text{comb}(t)\} = \frac{1}{T} \sum_{n=-\infty}^{\infty} \int_{-\infty}^{\infty} e^{-j(\omega - n\omega_s)t}\, dt = \frac{2\pi}{T} \sum_{n=-\infty}^{\infty} \delta(\omega - n\omega_s) = \frac{2\pi}{T} \text{COMB}_{\omega_s}(\omega)$$

1.4.1 $F(\omega) = \int\limits_{-\infty}^{\infty} f(t)e^{-\omega t}\,\mathrm{d}t = \lim\limits_{T\to 0} \sum\limits_{n=-\infty}^{\infty} \int\limits_{nT-T}^{nT} f(t)e^{-t\omega}\,\mathrm{d}t$

$\qquad \cong \sum\limits_{n=-\infty}^{\infty} f(nT)(nT - nT + T)e^{-\omega nT} = T\sum\limits_{n=-\infty}^{\infty} f(nT)e^{-\omega nT}$

1.5.1 Rewrite Equation 1.14 in the form:

$$f(-n) = \frac{1}{N}\sum\limits_{k=0}^{N-1} F(k)e^{j2\pi(-n)k/N}$$

Now, interchange the parameters n and k, which yields

$$f(-k) = \frac{1}{N}\sum\limits_{n=0}^{N-1} F(n)e^{-j2\pi kn/N}$$

This operation is the DFT of $F(\cdot)$ described in the time domain.

1.5.2 We write the IDFT in the form:

$$f(r) = \frac{1}{N}\sum\limits_{k=0}^{N-1} F(k)e^{j2\pi rk/N}$$

$$f(n-m) = \frac{1}{N}\sum\limits_{k=0}^{N-1} F(k)e^{j2\pi k(n-m)/N} = \left\{\frac{1}{N}\sum\limits_{k=0}^{N-1} F(k)e^{-j2\pi km}\right\}e^{j2\pi nk/N}$$

$$= \mathfrak{F}_D^{-1}\{F(k)e^{-j2\pi mk/N}\}$$

1.5.3 $F(k-m) = \sum\limits_{n=0}^{N-1} f(n)e^{-j2\pi(k-m)n/N} = \sum\limits_{n=0}^{N-1}\left[f(n)e^{j2\pi mn/N}\right]e^{-j2\pi kn/N}$

1.5.4 $y(n) = \sum\limits_{m=0}^{N-1} f(m)h(n-m) = \sum\limits_{m=0}^{N-1}\frac{1}{N}\sum\limits_{k=0}^{N-1} F(k)e^{j2\pi mk/N} \times \frac{1}{N}\sum\limits_{r=0}^{N-1} H(r)e^{j2\pi r(n-m)/N}$

$$= \frac{1}{N}\sum\limits_{k=0}^{N-1}\sum\limits_{r=0}^{N-1} F(k)H(r)e^{j2\pi rn/N}\left[\frac{1}{N}\sum\limits_{m=0}^{N-1} e^{j2\pi km/N}e^{-j2\pi mr/N}\right]$$

The expression in the bracket is equal to 1 if $k = r$ and 0 for $k \neq r$. To show this, use the finite geometric series formula. Hence, for $r = k$ in the second sum, we find the desired result:

$$y(n) = \sum_{m=0}^{N-1} f(m)h(n-m) = \frac{1}{N} \sum_{k=0}^{N-1} F(k)H(k)e^{j2\pi kn/N} = \mathfrak{F}_D^{-1}\{F(k)H(k)\}$$

1.5.7 Hint: In the continuous case the FT of 1 is $2\pi\delta(\omega)$.

1.7.1 Answer:

$$y(t) = e^{-t} - e^{-2t}, \quad t \geq 0$$

Appendix 1.1

Table of FT Pairs

$f(t) = \dfrac{1}{2\pi} \displaystyle\int_{-\infty}^{\infty} F(\omega)e^{j\omega t}\, d\omega$	$F(\omega) = \displaystyle\int_{-\infty}^{\infty} f(t)e^{-j\omega t}\, dts$
$f(t) = \begin{cases} 1, & \lvert t \rvert \leq a \\ 0, & \text{otherwise} \end{cases}$	$F(\omega) = 2\dfrac{\sin a\omega}{\omega}$
$f(t) = Ae^{-at}\, u(t)$	$F(\omega) = \dfrac{A}{a+j\omega}$
$f(t) = Ae^{-a\lvert t \rvert}$	$F(\omega) = \dfrac{2aA}{a^2 + \omega^2}$
$f(t) = \begin{cases} A(1-(\lvert t \rvert / a)), & \lvert t \rvert \leq a \\ 0, & \text{otherwise} \end{cases}$	$F(\omega) = Aa\left[\dfrac{\sin(a\omega/2)}{(a\omega/2)}\right]^2$
$f(t) = Ae^{-a^2 t^2}$	$F(\omega) = A\dfrac{\sqrt{\pi}}{a}e^{-(\omega/2a)^2}$
$f(t) = \begin{cases} A\cos\omega_0 t\, e^{-at}, & t \geq 0 \\ 0, & t < 0 \end{cases}$	$F(\omega) = A\dfrac{a+j\omega}{(a+j\omega)^2 + \omega_0^2}$
$f(t) = \begin{cases} A\sin\omega_0 t\, e^{-at}, & t \geq 0 \\ 0, & t < 0 \end{cases}$	$F(\omega) = \dfrac{A\omega_0}{(a+j\omega)^2 + \omega_0^2}$

Table of FT Pairs **(continued)**

$$f(t) = \frac{1}{2\pi}\int\limits_{-\infty}^{\infty} F(\omega)e^{j\omega t}\,d\omega \qquad\qquad F(\omega) = \int\limits_{-\infty}^{\infty} f(t)e^{-j\omega t}\,dts$$

$$f(t) = \begin{cases} A\cos\omega_0 t, & |t|\le a \\ 0, & \text{otherwise} \end{cases} \qquad F(\omega) = A\frac{\sin a(\omega-\omega_0)}{\omega-\omega_0} + A\frac{\sin a(\omega+\omega_0)}{\omega+\omega_0}$$

$$f(t) = A\delta(t) \qquad\qquad F(\omega) = A$$

$$f(t) = \begin{cases} A, & t>0 \\ 0, & \text{otherwise} \end{cases} \qquad F(\omega) = A\left[\pi\delta(\omega) - j\frac{1}{\omega}\right]$$

$$f(t) = \begin{cases} A, & t>0 \\ 0, & t=0 \\ -A, & t<0 \end{cases} \qquad F(\omega) = -j2A\frac{1}{\omega}$$

$$f(t) = A \qquad\qquad F(\omega) = 2\pi A\delta(\omega)$$

$$f(t) = A\cos\omega_0 t \qquad\qquad F(\omega) = \pi A[\delta(\omega-\omega_0) + \delta(\omega+\omega_0)]$$

$$f(t) = A\sum_{n=-\infty}^{\infty}\delta(t-nT) \qquad F(\omega) = \frac{2\pi A}{T}\sum_{n=-\infty}^{\infty}\delta\left(\omega - n\frac{2\pi}{T}\right)$$

$$f(t) = \frac{\sin at}{\pi t} \qquad\qquad F(\omega) = p_a(\omega)$$

$$f(t) = \frac{2\sin^2(at/2)}{\pi t^2} \qquad F(\omega) = \begin{cases} 1-(|\omega|/a), & |\omega|\le a \\ 0, & \text{otherwise} \end{cases}$$

Appendix 1.2

Summary of continuous time Fourier properties

Linearity: $af(t) + bh(t) \xleftrightarrow{\;\mathfrak{F}\;} aF(\omega) + bH(\omega)$

Time shifting: $f(t\pm t_0) \xleftrightarrow{\;\mathfrak{F}\;} e^{\pm j\omega t_0}F(\omega)$

Symmetry:
$$\begin{cases} F(t) \xleftrightarrow{\Im} 2\pi f(-\omega) \\ 1 \xleftrightarrow{\Im} 2\pi\delta(-\omega) = 2\pi\delta(\omega) \end{cases}$$

Time scaling:
$$f(at) \xleftrightarrow{\Im} \frac{1}{|a|} F\left(\frac{\omega}{a}\right)$$

Time reversal:
$$f(-t) \xleftrightarrow{\Im} F(-\omega)$$

Frequency shifting:
$$e^{\pm j\omega_0 t} f(t) \xleftrightarrow{\Im} F(\omega \mp \omega_0)$$

Modulation:
$$\begin{cases} f(t)\cos\omega_0 t \xleftrightarrow{\Im} \frac{1}{2}[F(\omega+\omega_0)+F(\omega-\omega_0)] \\ f(t)\sin\omega_0 t \xleftrightarrow{\Im} \frac{1}{2j}[F(\omega-\omega_0)-F(\omega+\omega_0)] \end{cases}$$

Time differentiation:
$$\frac{d^n f(t)}{dt^n} \xleftrightarrow{\Im} (j\omega)^n F(\omega)$$

Frequency differentiation:
$$\begin{cases} (-jt)f(t) \xleftrightarrow{\Im} \frac{dF(\omega)}{d\omega} \\ (-jt)^n f(t) \xleftrightarrow{\Im} \frac{d^n F(\omega)}{d\omega^n} \end{cases}$$

Time convolution:
$$f(t) * h(t) = \int_{-\infty}^{\infty} f(x)h(t-x)dx \xleftrightarrow{\Im} F(\omega)H(\omega)$$

Frequency convolution:
$$f(t)h(t) \xleftrightarrow{\Im} \frac{1}{2\pi} F(\omega) * H(\omega)$$
$$= \frac{1}{2\pi} \int_{-\infty}^{\infty} F(x)H(\omega-x)dx$$

Autocorrelation:
$$f(t) \odot f(t) = \int_{-\infty}^{\infty} f(x)f^*(x-t)dx \xleftrightarrow{\Im} F(\omega)F^*(\omega)$$
$$= |F(\omega)|^2$$

Central ordinate:
$$f(0) = \frac{1}{2\pi} \int_{-\infty}^{\infty} F(\omega)d\omega, \quad F(0) = \int_{-\infty}^{\infty} f(t)dt$$

Parseval's theorem:
$$E = \int_{-\infty}^{\infty} |f(t)|^2 dt, \quad E = \frac{1}{2\pi} \int_{-\infty}^{\infty} |F(\omega)|^2 d\omega$$

chapter two

Random variables, sequences, and stochastic processes

For most situations signals are not repeatable in a particular manner. For example, radar signals reflected from the ground, communication signals propagating through the atmospheric channel, the engine noise in speech transmission from the cockpit of an airplane, speckles in images, etc. These types of signals are defined by probabilistic characterization, and are treated by the theory of probability and statistics.

2.1 Random signals and distributions

Random signals can be described by precise mathematical analysis whose tools are contained in the theory of statistical analysis.

A discrete random signal $\{X(n)\}$ is a sequence of indexed random variables (rvs) assuming the values:

$$\{x(0) \quad x(1) \quad x(2) \quad ...\} \tag{2.1}$$

The random sequence with values $\{x(n)\}$ is discrete with respect to sampling index n. In our case, here, we will assume that the rv at any time n takes continuous values and, hence, it is a continuous rv at any time n. This type of sequence is also known as time series. In case we study a continuous random signal, we will assume that we sample it at high enough rate so that we construct a time series that is free of aliasing (see Section 1.3).

A particular rv continuous at time n, $X(n)$, is characterized by its probability density function (pdf) $f(x(n))$

$$\boxed{f(x(n)) = \frac{\partial F(x(n))}{\partial x(n)}} \tag{2.2}$$

and its cumulative density function (cdf) $F(x(n))$

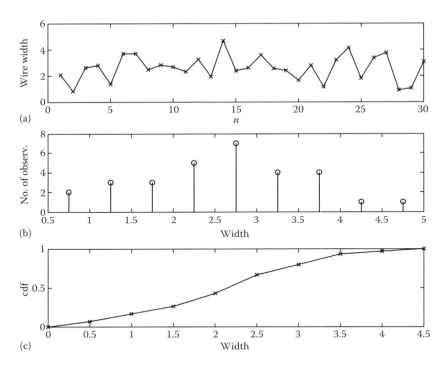

Figure 2.1

$$F(x(n)) = p\{X(n) \le x(n)\} = \int_{-\infty}^{x(n)} f(y(n)) \mathrm{d}y(n) \qquad (2.3)$$

The expression $p\{X(n) \le x(n)\}$ is interpreted as the probability that the rv $X(n)$ will take values less than or equal to $x(n)$ at time n. As the value of rv at time n approaches infinity, $F(x(n))$ approaches unity.

Figure 2.1a presents a time series indicating, for example, the thickness of a wire as it is measured at 30 instants of time during its production for quality assurance. Figure 2.1b shows the histogram, that is an approximation to the pdf of the rv, which in this case is the thickness of the wire. For example, the number of times we observed that the width falls between 2 and 2.5 is 5. Finally Figure 2.1c shows the cdf, where each step is proportional to the number of widths found within the range of width 0.5 divided by the total number of samples.

The empirical cdf of a rv, given the values $x_1(n)$, $x_2(n)$, ..., $x_N(n)$, is given by

$$\hat{F}(x \mid x_1, \ldots, x_N) = \frac{\text{Number of samples values } x_1, \ldots, x_N \text{ not greater than } x}{N} \qquad (2.4)$$

This function is a histogram of stairway type and its value at x (here the width) is the percentage of the points $x_1, ..., x_N$ that are not larger than x.

Similarly, the empirical pdf can be found based on the relation:

$$\hat{f}(x) \cong \frac{p\{x < X < x + \Delta x\}}{\Delta x}$$

for small Δx. Thus, we write

$$\hat{f}(x \mid x_1, ..., x_N) = \frac{\text{Number of the samples } x_1, ..., x_N \text{ in } [x, x + \Delta x)}{N \Delta x} \quad (2.5)$$

Similarly, the **multivariate** distributions are given by

$$F(x(n_1), ..., x(n_k)) = p\{X(n_1) \le x(n_1), ..., X(n_k) \le x(n_k)\}$$

$$f(x(n_1), ..., x(n_k)) = \frac{\partial^k F(x(n_1), ..., x(n_k))}{\partial x(n_1) \cdots \partial x(n_k)} \quad (2.6)$$

Note that here we have used a capital letter to indicate rvs. In general, we shall not keep this notation since it will be obvious from the context.

If, for example, we want to check the accuracy of reading a dial by a person, we will have two readings, one due to the person and another due to the instruments. A simultaneous plot of these two readings, each one associated with a different orthogonal axis, will produce a scattering diagram but with a linear dependence. The closer the points fall on a straight line the more reliable the person's readings are. This example presets a case of a **bivariate** distribution. Figure 2.2a shows a scatter plot of $x(n)$s versus $x(n-1)$s of the time series $\{x(n)\}$. Figure 2.2b shows a similar plot but for $x(n)$s versus $x(n-4)$. It is apparent that the $x(n)$s and $x(n-1)$s are more correlated (tend to be around a straight line) than the second case.

To obtain a formal definition of a discrete-time stochastic process, we consider an experiment with a finite or infinite number of unpredictable outcomes from a sample space, $S(z_1, z_2, ...)$, each one occurring with a probability $p(z_i)$. Next, by some rule we assign a deterministic sequence $x(n, z_i)$, $-\infty < n < \infty$, to each element z_i of the sample space. The sample space, the probabilities of each outcome, and the sequences constitute a **discrete-time stochastic process** or **random sequence**. From this definition, we obtain the following four interpretations:

- $x(n, z)$ is a rv if n is fixed and z is variable
- $x(n, z)$ is a sample sequence called realization if z is fixed and n is variable

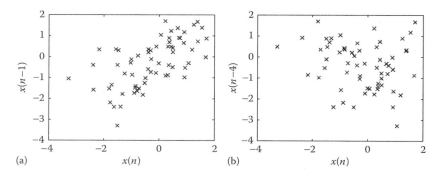

Figure 2.2

- $x(n, z)$ is a number if both n and z are fixed
- $x(n, z)$ is a stochastic process if both n and z are variables

Each time we run an experiment under identical conditions, we create a sequence of rvs {X(n)} which is known as a **realization** and constitutes **an event**. A realization is one member of a set called the **ensemble** of all possible results from the repetition of an experiment. Figure 2.3 shows a typical ensemble of realizations.

Book MATLAB® m-File: realizations

```
%Book MATLAB m-file: realizations
for n=1:4
    x(n,:)=rand(1,50)-0.5;%x=4×50 matrix with each row having
                    %zero mean;
end;
m=0:49;
for i=1:4
    subplot(4,1,i);stem(m,x(i,:),'k');%plots four rows of
                    %matrix x;
end;
xlabel('n');ylabel('x(n)')
```

Figure 2.3 shows four realizations of a stochastic process with zero mean value. With only slight modifications of the above script file we can produce any number of realizations.

Stationary and ergodic processes

It is seldom in practice that we will be able to create an ensemble of a random process with numerous realizations so that we can find some of its statistical

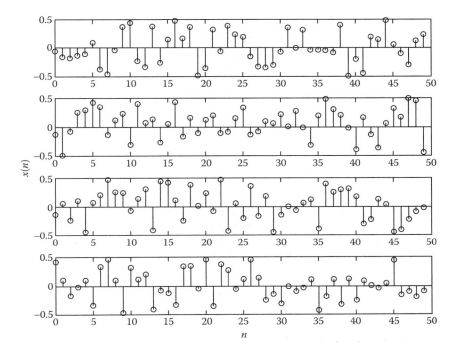

Figure 2.3

characteristics, e.g., mean value, variance, etc. To find these statistical quantities we need the pdf of the process, which, most of the times, is not possible to produce. Therefore, we will restrict our studies to processes which are easy to study and handle mathematically.

The process, which produces an ensemble of realizations and whose statistical characteristics do not change with time, is called **stationary**. For example, the pdf of the rvs $x(n)$ and $x(n + k)$ of the process $\{x(n)\}$ are the same independently of the values of n and k.

Since we will be unable to produce ensemble averages in practice, we are left with only one realization of the stochastic process. To overcome this difficulty, we assume that the process is **ergodic**. This characterization permits us to find the desired statistical characteristics of the process from only one realization at hand. We refer to those statistical values as **sample mean**, **sample variance**, etc. This assumes that the ergodicity is applicable to those statistical characteristics as well.

2.2   Averages

Mean value

The **mean** value or **expectation** value m_x at time n of a rv $x(n)$ having pdf $f(x(n))$ is given by

$$m_x(n) = E\{x(n)\} = \int\limits_{-\infty}^{\infty} x(n) f(x(n)) \mathrm{d}x(n) \qquad (2.7)$$

where $E\{\ \}$ stands for expectation operator. We can also use the ensemble of realizations to obtain the mean value using the **frequency-interpretation** formula:

$$m_x(n) = \lim_{N \to \infty} \left\{ \frac{1}{N} \sum_{i=1}^{N} x_i(n) \right\} \qquad (2.8)$$

where N is the number of realizations and $x_i(n)$ is the ith outcome at sample index n (or time n) of the ith realization. Depending on the type of rv, the mean value may or may not vary with time.

For an ergodic process, we find the sample mean **(estimator of the mean)** using the **time-average** formula (see Problem 2.2.1):

$$\hat{m} = \frac{1}{N} \sum_{n=0}^{N-1} x(n) \qquad (2.9)$$

It must be pointed out that the above estimate mean value is a rv depending on the number of terms of the sequence present. It turns out (see Problem 2.2.2) that the sample mean $\hat{m}$ is equal to the population mean m_x and, therefore, we call the sample mean an **unbiased** estimator.

Correlation

The **cross-correlation** between two random sequences is defined by

$$r_{xy}(m,n) = E\{x(m), y(n)\} = \iint x(m) y(n) f(x(m), y(n)) \mathrm{d}x(m) \mathrm{d}y(n) \qquad (2.10)$$

where the integrals are from $-\infty$ to ∞. If $x(n) = y(n)$, the correlation is known as the **autocorrelation**. Having an ensemble of realizations, the frequency interpretation of the autocorrelation function is found using the formula:

$$r_x(m,n) = \lim_{N \to \infty} \left\{ \frac{1}{N} \sum_{i=1}^{N} x_i(m) x_i(n) \right\} \qquad (2.11)$$

Note that we use one subscript for autocorrelation functions. In case of cross-correlation, we will use both subscripts.

Example 2.1

Using Figure 2.3, find the mean for $n = 10$ and the autocorrelation function for time difference of 5: $n = 20$ and 25.

Solution

The desired values are

$$m_{10} = \frac{1}{4}\sum_{i=1}^{4} x_i(10) = \frac{1}{4}(0.3 - 0.1 - 0.4 + 0.45) = 0.06$$

$$r_x(20,25) = \frac{1}{4}\sum_{i=1}^{4} x_i(25)x_i(30) = \frac{1}{4}[(0.1)(-0.35) + (-0.15)(-0.4)$$
$$+ (0.35)(0.25) + (0.2)(-0.1)]$$
$$= 0.032$$

Because the number of realizations is very small, both values found above are not as it was expected to be. However, their values are approximate what we were expected. ∎

Figure 2.4 shows the mean value at 50 individual times and the autocorrelation function for 50 differences (from 0 to 49) known as **lags**. These results were found using the MALAB function given below. Note that, as the number of realizations increases, the mean tends to zero and the autocorrelation tends to a delta function, as it should be. In this case, the rvs are independent, identically distributed (iid), and their pdf is Gaussian (white noise [WN], see Section 2.3.1).

Book MATLAB function for finding the mean and the autocorrelation function using the frequency interpretation approach: ssp_mean_autoc_ensemble(M,N)

```
%Book MATLAB function m-file: ssp_mean_autoc_ensemble
function[mx,rx]=ssp_mean_autoc_ensemble(M,N);
%N=number of time instances;easily modified for
%other pdf's;M=number of realizations;
x=randn(M,N);%randn=MATLAB function producing zero mean
            %Gaussian distributed white noise;x=MxN
            %matrix;
```

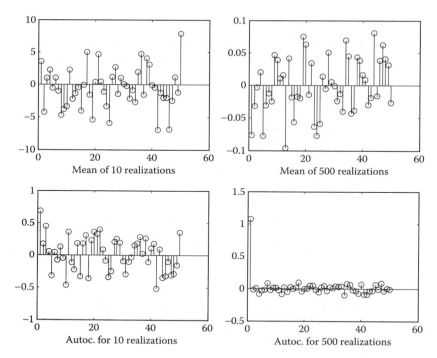

Figure 2.4

```
%sum(x,1)=MATLAB function that sums all the
%rows;
%sum(x,2)=MATLAB function that sums all the
%columns;
mx=sum(x,1)/M;
for i=1:N
    rx(i)=sum(x(:,1).*x(:,i))/M;
end;
```

At the command window and including the path that contains the above function, we write: > > [mx,rx] = ssp _ mean _ autoc _ ensemble(10, 50);subplot(2,1,1);stem(mx); subplot(2,1,2);stem(rx);. Next, we introduce the values $M = 500$ and $N = 50$ and thus producing the other two plots.

More will be discussed about the correlation function of sampled samples in Chapter 3.

Covariance

The covariance of a random sequence is defined by

$$c_x(m,n) = E\{(x(m) - m_m)(x(n) - m_n)\} = E\{x(m)x(n)\} - m_m m_n$$
$$= r_x(m,n) - m_m m_n \qquad (2.12)$$

The variance is found by setting $m = n$ in Equation 2.12. Thus,

$$c_x(n,n) = \sigma_n^2 = E\{(x(n) - m_n)^2\} = E\{x^2(n)\} - m_n^2 \qquad (2.13)$$

If the mean value is 0, then the variance and the correlation function at zero sift are identical.

$$c_x(n,n) = \sigma_n^2 = E\{x^2(n)\} = r_x(n,n) \qquad (2.14)$$

The estimator for the biased variance is given by

$$\hat{\sigma}^2 = \frac{1}{N}\sum_{n=1}^{N}(x(n) - \hat{m}_n)^2 \qquad (2.15)$$

and for the unbiased case we divide by $N - 1$. The above variance can be found from the Book MATLAB function **ssp_sample_biased_autoc** (see Chapter 3 for details) at 0 lag. The reader can also use the MATLAB functions **std**(data vector) = standard deviation and **var**(data vector) = variance.

Example 2.2

With the help of MATLAB, we obtain: $x = 2.5$*randn (1, 10) + 6.8 or $x = [7.5360\ 3.4595\ 8.5858\ 10.8589\ 5.0706\ 8.9450\ 9.9350\ 2.8157\ 3.1976\ 8.2279]$, $\hat{m} = (1/10)\text{sum}(x) = 6.8632$, variance $\hat{\sigma}^2$ = vari is found using the following MATLAB program: $n = 1:10$; for $n = 1:10$; $d(n) = (x(n) - 6.8632)^2$; vari = sum(d)/10; end; hence vari = 7.9636 and standard deviation is sqrt(7.9636) = $\hat{\sigma}$ = 2.8220. ■

Independent and uncorrelated rvs

If the joint pdf of two rvs can be separated into two pdfs, $f_{x,y}(m, n) = f_x(m)f_y(n)$, then the rvs are statistically **independent**. Hence,

$$E\{x(m)x(n)\} = E\{x(m)\}E\{x(n)\} = m_x(m)m_x(n) \qquad (2.16)$$

The above equation is necessary and sufficient condition for the two rvs $x(m)$, $x(n)$ to be uncorrelated. Note that independent rvs are always uncorrelated. However, the converse is not necessarily true. If the mean value of any two uncorrelated rvs is 0, then the rvs are called **orthogonal**. In general, two rvs are called orthogonal if their correlation is 0.

2.3   Stationary processes

For a **wide-sense (or weakly) stationary** (WSS) process, the cdf satisfies the relationship

$$F(x(m), x(n)) = F(x(m+k), x(n+k)) \tag{2.17}$$

for any m, n, and k. The above relationship also applies for all statistical characteristics such as mean value, variance, correlation, etc. If the above relationship is true for any number of rvs of the time series, then the process is known as strictly stationary process.

The basic **properties** of a wide-sense real stationary process are (see Problem 2.3.1):

$$m_x(n) = m = \text{constant} \tag{2.18a}$$

$$r_x(m, n) = r_x(m - n) \tag{2.18b}$$

$$r_x(k) = r_x(-k) \tag{2.18c}$$

$$r_x(0) \geq r_x(k) \tag{2.18d}$$

Autocorrelation matrix

If $\mathbf{x} = [x(0)\, x(1) \cdots x(p)]^T$ is a vector representing a finite random sequence, then the autocorrelation matrix is given by

$$
\mathbf{R}_x = E\{\mathbf{x}\mathbf{x}^T\} =
\begin{bmatrix}
E\{x(0)x(0)\} & E\{x(0)x(1)\} & \cdots & E\{x(0)x(p)\} \\
E(x(1)x(0)\} & E\{x(1)x(1)\} & \cdots & E\{x(1)x(p)\} \\
\vdots & \vdots & \vdots & \vdots \\
E\{x(p)x(0)\} & E\{x(p)x(1)\} & \cdots & E\{x(p)x(p)\}
\end{bmatrix}
$$

$$
=
\begin{bmatrix}
r_x(0) & r_x(-1) & \cdots & r_x(-p) \\
r_x(1) & r_x(0) & \cdots & r_x(-p+1) \\
\vdots & \vdots & \vdots & \vdots \\
r_x(p) & r_x(p-1) & \cdots & r_x(0)
\end{bmatrix}
\tag{2.19}
$$

However, in practical applications we will have a single sample at hand and this will permit us to find only the estimate correlation matrix $\hat{\mathbf{R}}_x$. Since, in general and in this chapter, we will have single realizations to work with, we will not explicitly indicate the estimate with the over bar.

Example 2.3

(a) Find the biased autocorrelation function with lag time up to 20 out of a sequence of 40 terms, which is a realization of rvs having Gaussian distribution with zero mean value; (b) create a 4×4 autocorrelation matrix.

Solution

The Book MATLAB function ex2_3_1 produces Figure 2.5. To find the correlation matrix from the autocorrelation function we use the following MATLAB function: $\mathbf{R_x} = \textbf{toeplitz}(\mathbf{r_x}(\textbf{1,1: 4}))$. In this example, a 4×4 matrix produced from the first four columns of correlation row vector $\mathbf{r_x}$ is

$$\mathbf{R}_x = \begin{bmatrix} 2.0817 & 0.3909 & 0.1342 & -0.0994 \\ 0.3909 & 2.0817 & 0.3909 & 0.1342 \\ 0.1342 & 0.3909 & 2.0817 & 0.3909 \\ -0.0994 & 0.1342 & 0.3909 & 2.0817 \end{bmatrix}$$ ∎

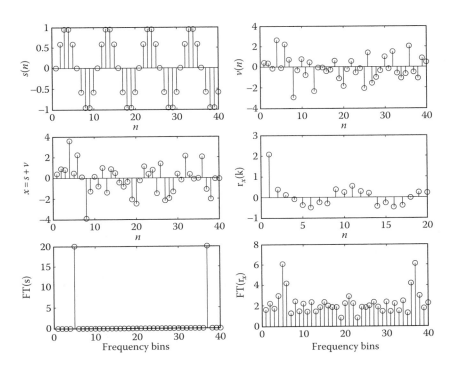

Figure 2.5

Book MATLAB m-File: ex2_3_1

```
%ex2_3_1 m-file
n=0:39;
s=sin(.2*pi*n);
v=randn(1,40); % randn=MATLAB function producing white
               % Gaussian
               % distributed rv's;
x=s+v;
rx=sasamplebiasedautoc(x,20);% Book MATLAB function
                             % creating the
                             % autocorrelation of x;
fts=fft(s,40); % fft=MATLAB function executing the fast
               % Fourier
               %transform;
ftrx=fft(rx,40);
subplot(3,2,1);stem(s, 'k');
xlabel('n');ylabel('s(n)');
subplot(3,2,2);stem(v,'k');
xlabel('n');ylabel('v(n)');
subplot(3,2,3);stem(x,'k');
xlabel('n');ylabel('x=s+v');
subplot(3,2,4);stem(rx,'k');
xlabel('lag number');ylabel('r_x');
subplot(3,2,5);stem(abs(fts),'k')
xlabel('freq. bins');ylabel('FT(s)');
subplot(3,2,6);stem(abs(ftrx),'k');
xlabel('freq. bins');ylabel('FT(r_x)');
```

Note: *If we have a row vector x and we need to create a vector y with elements of x from k to m only, we write: y = x(1, k:m). If x is a column vector, we write: y = x(k:m, 1).*

Example 2.4

Let $\{v(n)\}$ be a zero-mean, uncorrelated Gaussian random sequence with variance $\sigma_v^2(n) = \sigma^2 = \text{constant}$. (a) Characterize the random sequence $\{v(n)\}$, and (b) determine the mean and the autocorrelation of the sequence $\{x(n)\}$ if $x(n) = v(n) + av(n-1)$, in the range $-\infty < n < \infty$, where a is a constant.

Solution

(a) The variance of $\{v(n)\}$ is constant and, hence, is independent of the time, n. Since $\{v(n)\}$ is an uncorrelated sequence it is also independent due to the fact that it is a Gaussian sequence. From Equation 2.12, we obtain $c_v(l,n) = r_v(l,n) - m_l m_n = r_v(l,n)$ and $\sigma^2 = r_v(n,n) =$ constant. Hence, $r_v(l,n) = \sigma^2 \delta(l-n)$, which implies that $\{v(n)\}$ is a WSS process.

(b) $E\{x(n)\} = 0$ since $E\{v(n)\} = E\{v(n-1)\} = 0$. Hence,

$$
\begin{aligned}
r_x(l,n) &= E\{[v(l) + av(l-1)][v(n) + av(n-1)]\} \\
&= E\{v(l)v(n)\} + aE\{v(l-1)v(n)\} + aE\{v(l)v(n-1)\} \\
&\quad + a^2 E\{v(l-1)v(n-1)\} \\
&= r_v(l,n) + ar_v(l-1,n) + ar_v(l,n-1) + a^2 r_v(l-1,n-1) \\
&= \sigma^2 \delta(l-n) + a\sigma^2 \delta(l-n+1) + a^2\sigma^2 \delta(l-n) \\
&= (1+a^2)\sigma^2 \delta(r) + a\sigma^2 \delta(r+1) + a\sigma^2 \delta(r-1), \quad l-n = r
\end{aligned}
$$

Since the mean of $\{x(n)\}$ is 0, a constant, and its autocorrelation is a function of the lag factor $r = l - n$, $\{x(n)\}$ is a WSS process. ∎

Purely random process (WN)

A discrete process is a **purely random process** if the rvs $\{x(n)\}$ are a sequence of mutually independent and **identically distributed** (id) variables. Since the mean and cov($x(m), x(m-k)$) do not depend on time, the process is WSS. This process is also known as WN and is given by

$$
\begin{aligned}
c_x(k) &= \begin{cases} E\{(x(m) - m_x)(x(m-k) - m_x)\} = \sigma_x^2 \delta(k) - m_x^2, & k = 0 \\ 0, & k = \pm 1, \pm 2, \ldots \end{cases} \\
\delta(k) &= \begin{cases} 1, & k = 0 \\ 0, & k \neq 0 \end{cases} \\
r_x(k) &= c_x(k), \quad m_x = 0
\end{aligned}
\tag{2.20}
$$

Since the process is white, which implies that for k different than zero $E\{(x(i) - m_x)(x(i-k) - m_x)\} = [E\{x(i)\} - m_x][E\{x(i-k)\} - m_x] = 0$

Random walk (RW)

Let $\{x(n)\}$ be a purely random process (iid rvs) with mean m_x and variance σ_x^2. A process $\{y(n)\}$ is a RW if

$$
y(n) = y(n-1) + x(n), \quad y(0) = 0
\tag{2.21}
$$

Therefore, the process takes the form:

$$y(n) = \sum_{i=1}^{n} x(i) \tag{2.22}$$

which is found from Equation 2.21 by recursive substitution ($y(1) = y(0) + x(0)$; $y(2) = y(1) + x(1) = 0 + x(0) + x(2)$; $y(3) = y(2) + x(1) = x(0) + x(1) + x(2)$; etc.).

The mean is found to be $E\{y(n)\} = nm_x$ and the $\mathrm{cov}(y(n), y(n)) = n[E\{x^2\} - m_x^2] = n\sigma_x^2$ (see Problem 2.3.7). It is interesting to note that the difference $x(n) = y(n) - y(n-1)$ is purely random and, hence, stationary.

Moving average (MA) process

If $\{x(n)\}$ is purely a random process (iid) with mean 0 and variance σ_x^2, then the process

$$\boxed{y(n) = b(0)x(n) + b(1)x(n-1) + \cdots + b(q)x(n-q)} \tag{2.23}$$

is known as the MV process of order q. The xs can be scaled so that we can also set $b(0) = 1$.

Since $x(n)$s are iid with zero mean value, it is easy to show that (see Problem 2.3.6):

$$E\{y(n)\} \triangleq m_y = 0, \qquad \mathrm{var}(y(n)) \triangleq E\{y^2(0)\} = r_y(0) \triangleq \sigma_y^2$$

$$= \sigma_x^2 \sum_{i=0}^{q} b^2(i), \quad k = 0 \tag{2.24}$$

For lag values $k > 0$ and for shift $k = 1$, we proceed as follows:

$$r_y(1) = E\{[b(0)x(n) + b(1)x(n-1) + b(2)x(n-2) + b(3)x(n-3)]$$
$$[b(0)x(n-1) + b(1)x(n-2) + b(2)x(n-3) + b(3)x(n-4)]\}$$
$$= b(1)b(0)E\{x^2(n-1)\} + b(2)b(1)E\{x^2(n-2)\} + b(3)b(2)E\{x^2(n-3)\}$$
$$= \sigma_x^2[b(1)b(0) + b(2)b(1) + b(3)b(2)] = \sigma_x^2 \sum_{m=1}^{3} b(m)b(m-1)$$

Terms involving xs at different times have been neglected because their product has expectation zero due to the fact that xs are iid and have zero mean value. For $n > q$, they are not xs with common times and thus the correlation is 0. If we now set the shift by q, then the above equation takes the general form:

$$r_y(k) = \sum_{m=k}^{q} b(m)b(m-k), \quad k = 0, 1, 2, \ldots, q \qquad (2.25)$$

By including $k = 0$, Equation 2.25 includes the result of Equation 2.24.

The process is WSS since the correlation is independent of time n and its mean value is constant. If the rvs $x(n)$s are normally distributed, then the $y(n)$s are also normally distributed and the process is completely determined by the mean and auto-covariance function. Therefore, the process under the above assumption is strictly stationary.

Autoregressive (AR) process

A process $y(n)$ is AR of order p if it is represented by the difference equation:

$$y(n) = -a(1)y(n-1) - a(2)y(n-2) - \cdots - a(p)y(n-p) + x(n)$$
$$= -\sum_{i=1}^{p} a(i)y(n-i) + x(n) \qquad (2.26)$$

where $x(n)$ is a purely random process with zero mean and variance σ_x^2.

The first-order process is called **Markov process**, after the Russian mathematician A. Markov, and is defined by the first-order difference equation:

$$y(n) = -a(1)y(n-1) + x(n) \qquad (2.27)$$

Multiply Equation 2.26 by $y(n-k)$ and take the ensemble average both sides to obtain

$$E\{y(n)y(n-k)\} = -\sum_{i=1}^{p} a(i)E\{y(n-i)y(n-k)\} + b(0)E\{x(n)y(n-k)\} \qquad (2.28)$$

or

$$r_y(k) = -\sum_{i=1}^{p} a(i)r_y(k-i) + b(0)r_{xy}(k) \qquad (2.29)$$

Since the input to the system is only the WN $x(n)$, the cross-correlation $r_{xy}()$ is valid only for $k = 0$, since the output is also white.

For any input, the output of any linear time-invariant (LTI) system is given by

$$y(n) = \sum_{i=0}^{\infty} h(i)x(n-i) \qquad (2.30)$$

Multiply both sides of Equation 2.30 and take the expectation of both sides of the equation:

$$E\{x(n)y(n-k)\} \triangleq r_{xy}(k) = \sum_{i=0}^{\infty} h(i)E\{x(n)x(n-k-i)\} = \sum_{i=0}^{\infty} h(i)r_x(k+i)$$

Since the input to the system is only at time n, the cross-correlation has only value when $k = 0$. Hence, from the above equation we obtain

$$r_{xy}(0) = h(0)r_x(0) = h(0)\sigma_x^2 \tag{2.31}$$

Since any general system can be presented in the form:

$$H(z) = \frac{b(0)+b'(1)z^{-1}+\cdots+b'(q)z^{-q}}{a(0)+a'(1)z^{-1}+\cdots+a'(p)z^{-p}} = \frac{b(0)}{a(0)} \frac{1+b(1)z^{-1}+\cdots+b(q)z^{-q}}{1+a(1)z^{-1}+\cdots+a(p)z^{-p}} \tag{2.32}$$

we can always set $a(0) = b(0) = 1$ without any loss of generality. Using the limiting property of the z-transform, we obtain

$$h(0) = \lim_{z \to \infty} H(z) = 1$$

Therefore, Equation 2.29 becomes

$$r_y(k) = \begin{cases} -\sum_{i=1}^{p} a(i)r_y(-i)+\sigma_x^2, & k=0 \\ -\sum_{i=1}^{p} a(i)r_y(k-i), & k>0 \end{cases} \tag{2.33}$$

Example 2.5

Find the output of the first-order AR system $y(n) = -0.85y(n-1) + x(n)$ when the input is WN.

Solution

The following MATLAB program produces Figure 2.6.

Book MATLAB m-File: ex2_3_3

```
%m-file: ex2_3_3
y(1)=0;
for n=2:50
    v(n)=rand;
    y(n)=-0.85*y(n-1)+v(n);
end;
```

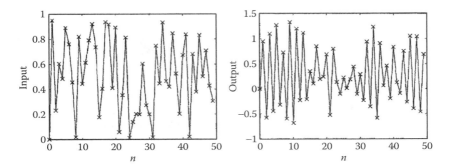

Figure 2.6

```
m=0:49;
subplot(2,2,1);plot(m,v,'kx-');xlabel('n');ylabel('Input');
subplot(2,2,2);plot(m,y,'kx-');xlabel('n');ylabel('Output');
```

Yule–Walker equations

Equation 2.33 can be written in the following matrix form:

$$\begin{bmatrix} r_y(0) & r_y(-1) & \cdots & r_y(-p) \\ r_y(1) & r_y(0) & \cdots & r_y(-p+1) \\ \vdots & \vdots & & \vdots \\ r_y(p) & r_y(p-1) & \cdots & r_y(0) \end{bmatrix} \begin{bmatrix} 1 \\ a(1) \\ \vdots \\ a(p) \end{bmatrix} = \sigma_x^2 \begin{bmatrix} 1 \\ 0 \\ \vdots \\ 0 \end{bmatrix} \qquad (2.34)$$

2.4   Wiener–Khinchin relations

For a WSS process, the correlation function asymptotically goes to zero and, therefore, we can find its spectrum using the discrete-time Fourier transform (DTFT). Hence, the **power spectrum** is given by

$$S_x(e^{j\omega}) = \sum_{k=-\infty}^{\infty} r_x(k)e^{-j\omega k} \qquad (2.35)$$

This function is periodic with period 2π since $\exp(-jk(\omega + 2\pi)) = \exp(-jk\omega)$). Given the power spectral density of a sequence, the autocorrelation of the sequence is given by the relation:

$$r_x(k) = \frac{1}{2\pi} \int_{-\pi}^{\pi} S_x(e^{j\omega})e^{j\omega k} \, d\omega \qquad (2.36)$$

For real process, $r_x(k) = r_x(-k)$ (symmetric function) and as a consequence the power spectrum is an **even** function. Furthermore, the power spectrum of

WSS process is also nonnegative. These two assertions are given below in the form of mathematical relations:

$$S_x(e^{j\omega}) = S_x(e^{-j\omega}) = S_x^*(e^{j\omega})$$
$$S_x(e^{j\omega}) \geq 0 \tag{2.37}$$

The positive property of the power spectrum will be apparent when we define a second form of the power spectrum of random processes.

Example 2.6

Find the power spectra density of the sequence $x(n) =$ sin(0.1*2*pi*n) + 1.5*randn(1,32) with $n = [0\ 1\ 2 \ldots 31]$.

Solution

The following Book MATLAB m-file produces Figure 2.7.

Book MATLAB m-File: ex2_4_1

```
%Book MATLAB m-file: ex2_4_1
n=0:31;s=sin(0.1*pi*n);v=randn(1,32);%white Gaussian
                %noise;
x=s+v;
r=xcorr(x,'biased');%the biased autocorrelation function
                %is divided by N=length(x);
fs=fft(s,32);fr=fft(r,32);
subplot(3,2,1);stem(n,s,'k');xlabel('n');ylabel('s(n)');
subplot(3,2,2);stem(n,v,'k');xlabel('n');ylabel('v(n)');
subplot(3,2,3);stem(n,x,'k');xlabel('n');ylabel('x(n)=...
  s(n)+v(n)');
subplot(3,2,4);stem(n,r(1,32:63),'k');xlabel('k, time lag');...
ylabel('r_x(k)');
subplot(3,2,5);stem(n,abs(fs),'k');xlabel('freq. bins');...
ylabel('S_s(e^{j\omega})');
subplot(3,2,6);stem(n,abs(fr),'k');xlabel('freq. bins');...
ylabel('S_x(e^{j\omega})');
```

If we set $z = e^{j\omega}$ in Equation 2.35, we obtain the z-transform of the correlation function instead of the DTFT. Hence,

$$\boxed{S_x(z) = \sum_{k=-\infty}^{\infty} r_x(k)z^{-k}} \tag{2.38}$$

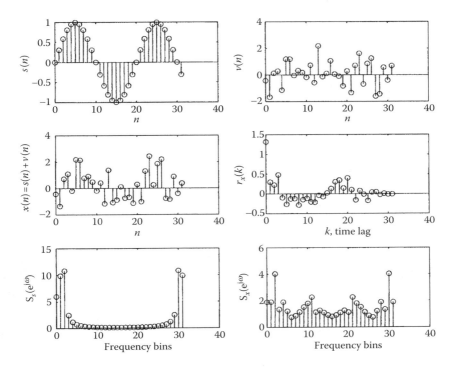

Figure 2.7

The total power of a zero mean WSS random process is proportional to the area under the power density curve and is given by the following equation:

$$r_x(0) = E\{x^2(n)\} = \frac{1}{2\pi} \int_{-\pi}^{\pi} S_x(e^{j\omega}) d\omega \qquad (2.39)$$

Example 2.7

Find the power spectrum of a WN process $\{x(n)\}$.

Solution

Since the process is iid, its correlation function is $r_x(k) = \sigma_x^2 \delta(k)$ and, hence, the power spectrum is

$$S_x(e^{j\omega}) = \sum_{k=-\infty}^{\infty} \sigma_x^2 \delta(k) e^{j\omega k} = \sigma_x^2 \qquad (2.40)$$

■

Apply Equation 2.35 for the correlation function:

$$r_x(k) = a^{|k|}, \quad -\infty < k < \infty, \quad 0 < a < 1 \tag{2.41}$$

to obtain the following power spectra density (see Problem 2.4.1):

$$S_x(e^{j\omega}) = \frac{1-a^2}{1-2a\cos\omega+a^2} \tag{2.42}$$

2.5 Filtering random processes

LTI filters (systems) are used in many signal processing applications. Since the input signals of these filters are usually random processes, we need to determine how the statistics of these signals are modified as a result of filtering.

Let $x(n)$, $y(n)$, and $h(n)$ be the filter input, filter output, and the filter impulse response, respectively. It can be shown (see Problem 2.5.1) that if $x(n)$ is WSS process, then the filter output autocorrelation $r_y(k)$ is related to the filter input autocorrelation $r_x(k)$ as follows:

$$
\begin{array}{|l|}
\hline
\\
r_y(k) = \displaystyle\sum_{l=-\infty}^{\infty} \sum_{m=-\infty}^{\infty} h(l)r_x(m-l+k)h(m) \\[2mm]
\quad = r_x(k)*h(k)*h(-k) = r_x(k)*r_h(k) \\[2mm]
h(k)*h(-k) = \text{Convolution between } h(k) \text{ and its reflected form } h(-k) \\[2mm]
\quad = \text{Autocorrelation of } h(k) \\
\hline
\end{array} \tag{2.43}
$$

The variance, $\sigma_y^2 = r_y(0)$, of the output of the system is found from the above equation by setting $k = 0$. Hence,

$$\sigma_y^2 = r_y(0) = \sum_{l=-\infty}^{\infty} \sum_{m=-\infty}^{\infty} h(l)r_x(m-l)h(m) \tag{2.44}$$

If $h(n)$ is 0 outside the interval $[0, N-1]$, then the variance of $y(n)$ (power) can be expressed as follows:

$$\sigma_y^2 = E\{y^2(n)\} = \mathbf{h'R_x h} \tag{2.45}$$

The prime means the transpose of the vector. The reader will easily show that for $N = 2$, Equations 2.44 and 2.45 give identical results.

If we take the z-transform of Equation 2.43, and taking into consideration the property for convolution, we obtain the power spectra of the output of the system to be

$$\mathcal{Z}\{r_y(k)\} = \mathcal{Z}\{r_x(k)\}\,\mathcal{Z}\{h(k)\}\,\mathcal{Z}\{h(-k)\} \tag{2.46}$$

$$S_y(z) = S_x(z)H(z)H(z^{-1}) \tag{2.47a}$$

$$S_y(e^{j\omega}) = S_x(e^{j\omega})H(e^{j\omega})H(e^{-j\omega}) \tag{2.47b}$$

$$\boxed{S_y(e^{j\omega}) = S_x(e^{j\omega})\,|\,H(e^{j\omega})\,|^2} \tag{2.47c}$$

Example 2.8

Two systems have the transfer functions: (a) $H_1(z) = 1/(1 - 0.2z^{-1})$ and (b) $H_2(z) = 1 + 0.4z^{-1}$. If the input to both systems is a WN $v(n)$ with variance σ_v^2 find the output spectrums.

Solution

From Equations 2.40 and 2.47, we obtain

(a) $S_y(z) = \sigma_v^2 H(z)H(z^{-1}) = \sigma_v^2 \dfrac{1}{1-0.2z^{-1}}\dfrac{1}{1-0.2z}$

or

$$S_y(e^{j\omega}) = \frac{\sigma_v^2}{1-0.4\cos\omega+0.04}$$

(b) $S_y(z) = \sigma_v^2 H(z)H(z^{-1}) = \sigma_v^2(1+0.4z^{-1})(1+0.4z)$

or

$$S_y(e^{j\omega}) = \sigma_v^2(1.16+0.8\cos\omega)$$ ■

2.6   *Probability density functions*

Some of the important properties of the pdfs are

1. The pdf is nonnegative

$f(x(n)) \geq 0, \quad \text{for all } x$

2. Integrating pdf yields cdf:

$$F(x(n)) = \int_{-\infty}^{x(n)} f(y(n))dy(n) = \text{Area under pdf over the interval}(-\infty, x] \tag{2.48}$$

3. Normalization property

$$\int_{-\infty}^{\infty} f(x(n))\,dx(n) = F(\infty) = 1 \tag{2.49}$$

4. Area under $f(x(n))$ over the interval $(x_1(n), x_2(n))$ is equal to $p\{x_1(n) < X(n) \le x_2(n)\}$,

$$p\{x_1(n) < X(n) \le x_2(n)\} = \int_{x_1(n)}^{x_2(n)} f(x(n))\,dx(n) \tag{2.50}$$

2.6.1   Uniform distribution

A WSS discrete random sequence that satisfies the relation:

$$f(x(0), x(1), \ldots) = f(x(0))f(x(1))f(x(2))\ldots \tag{2.51}$$

is a purely random sequence whose elements $x(n)$ are statistically independent and iid. A rv $x(n)$ at time n has a uniform distribution if its pdf is of the form:

$$f(x) = \begin{cases} \dfrac{1}{b-a}, & a \le x \le b \\ 0, & \text{elsewhere} \end{cases} \tag{2.52}$$

The MATLAB function **rand(1,1000)**, for example, will provide a random sequence of 1000 elements whose sample mean value is 0.5 and its pdf is uniform from 0 to 1 with amplitude 1. To produce the pdf, we can use the MTLAB function **hist(x,20)**. This means that we want to produce the pdf dividing the range of values, in this case from 0 to 1, of the random vector **x** in 20 equal steps. They are two more useful functions of MATLAB that give the mean value, **mean(x)**, and the variance, **var(x)**. We can also find the standard deviation using the MATLAB function **std(x)**, which is equal to the square root of the variance.

Note that

$$\int_{-\infty}^{\infty} f(x)\,dx = \int_{a}^{b} (1/(b-a))\,dx = 1$$

and

$$F(x) = \int_{-\infty}^{x} f(y)\,dy = \begin{cases} 0, & x < 0 \\ \int_{a}^{x} (1/(b-a))\,dy = ((x-a)/(b-a)), & a \le x \le b \\ 1, & x > b \end{cases} \tag{2.53}$$

Other statistical characteristics of the uniform distribution:
   Range: $a \leq x \leq b$
   Mean: $(a + b)/2$
   Variance: $(b - a)^2/12$

2.6.2   Gaussian (normal) distribution

The pdf of a Gaussian rv $x(n)$ at time n with mean value m_x and variance σ_x^2 is

$$f(x) = \frac{1}{\sqrt{2\pi\sigma_x^2}} e^{-(x-m_x)^2/2\sigma_x^2} = N(m_x, \sigma_x^2) \qquad (2.54)$$

Example 2.9

Find the joint pdf of a white Gaussian noise (WGN) with N elements each one having zero mean value and the same variance.

Solution

The joint pdf is

$$f(x(1), x(2), \ldots, x(N)) = f_1(x(1))f_2(x(2)) \ldots f_N(x(N))$$

$$= \frac{1}{(2\pi)^{N/2}\sigma_x^N} \exp\left[-\frac{1}{2\sigma_x^2}\sum_{k=1}^{N} x^2(k)\right] \qquad (2.55) \quad \blacksquare$$

A discrete-time random process $\{x(n)\}$ is said to be Gaussian if every finite collection of samples of $x(n)$ are jointly Gaussian. A Gaussian random process has the following properties: (a) is completely defined by its mean vector and covariance matrix; (b) any linear operation on the time variables produces another Gaussian random process; (c) all higher moments can be expressed by the first and second moments of the distribution (mean, covariance); and (d) WN is necessarily generated by iid samples (independence implies uncorrelated rvs and vice versa).

Example 2.10

If the pdf $f(x)$ is $N(2.5,4)$, find the $p\{x \leq 6.5\}$.

Solution

$$p\{x \leq 6.5\} = \int_{-\infty}^{6.5} (1/\sqrt{2\pi})(1/2)e^{-(x-2.5)^2/8}\,dx \quad (1). \text{ Set in } (1)\ y =$$
$(x - 2.5)/2$ to obtain: for $x = -\infty \Rightarrow y = -\infty$, for $x = 6.5 \Rightarrow y = 2$ and $dx = 2dy$. Hence (1) becomes $P\{x \leq 6.5\} = P\{y \leq 2\} =$
$\left(1/\sqrt{2\pi}\right)\int_{-\infty}^{0} e^{-y^2/2}\,dy + (1/2\pi)\int_{0}^{2} e^{-y^2/2}\,dy = 0.5 + \text{erf}(2)$. The

error function is tabulated, but MATLAB can give very accurate values by introducing sampling values <0.0001. For this sampling value, MATLAB gives the results $erf(2) = (1/sqrt(2*pi))*0.0001*sum(-y.^2/2) = 0.47727$, where the vector **y** is given by **y** = 0:0.0001:2. Tables give $erf(2) = 0.47724$. Therefore, $p\{x\leq6.5\} = 0.97727$.   ■

To produce a WGN with mean zero value and unit variance, $N(0,1)$, the following MATLAB function can be used

```
x= randn(1,N); % x is a vector with N elements of WGN
               % type with zero mean and unit variance;
```

MATLAB function: hist(x,b)

```
hist(x,b); % x is a sample random vector from a
           % particular distribution;
           % b is the number of ranges we are wishing to
           % split the range
           % of the values of x from its minimum to its
           % maximum;
```

Transformation to $N(m,va)$ from the $N(0,1)$ of the rv x

In case it is desired to have a WGN with mean m and variance va, we use the following transformation of the vector **x**.

$z = va*x + m$; % the variance of the time series z is va², and its mean value
                % is equal to m;
                % the variance of x is 1 and the mean value is 0;

Algorithm

1. Generate two independent rvs u_1 and u_2 from a uniform distribution (0, 1)
2. $x_1 = (-2 \ln(u_1))^{1/2} \cos(2\pi u_2)$ (or $x_2 = (-2 \ln(u_1))^{1/2} \sin(2\pi u_2)$)
3. Keep x_1 or x_2

Book MATLAB function: sspnormalpdf(m,s,N)

```
function[x]=sspnormalpdf(m,s,N)
% [x]=sspnormalpdf(m,s,N);N=number of elements in x;
%s=standard deviation; m=mean value;
for i=1:N
    r1=rand;
```

```
    r2=rand;
    z(i)=sqrt(-2*log(r1))*cos(2*pi*r2);
end;
x=s*z+m;
```

Book MATLAB function: sspnormalpdf1(m,s,N)

```
function[x]=sspnormalpdf1(m,s,N);
z=sqrt(-2*log(rand(1,N))).*cos(2*pi*rand(1,N));
x=s*z+m;
```

We can also use a Monte Carlo approach to obtain the normal distributed rvs. The following program does this.

Book MATLAB function: sspmontecarlonormalpdf(m,s,N)

```
function[y]=sspmontecarlonormalpdf(m,s,N,M)
%m=mean value;s=standard deviation;
%N=number of variables;M=number of sumable normal
%variables;
for n=1:N
    x(n)=sum(randn(1,M))/sqrt(M);
end;
y=m+s*x;
```

The above Monte Carlo processing was based on the **central limit theorem,** which states:

For N independent rvs $X_1, X_2, \ldots, X_N$ with mean m_i and variance σ_i^2, respectively,

$$y_N = \frac{\sum_{i=1}^{N}(x_i - m_i)}{\sqrt{\sum_{i=1}^{N}\sigma_i^2}}, \quad \lim N \to \infty \Rightarrow y \equiv N(0,1) \tag{2.56}$$

The central limit theorem has the following interpretation:

> The properly normalized sum of many uniformly small and negligible independent rvs tends to be a standard normal (Gaussian) rv. If a random phenomenon is the cumulative effect of many uniformly small sources of uncertainty, it can be reasonably modeled as normal rv.

If the rv x is $N(m_x, \sigma_x^2)$ with $\sigma_x^2 > 0$, then the rv $w = (x - m_x)/\sigma_x$ is $N(0,1)$ (see Problem 2.5.2).

Other statistical characteristics of the normal distribution:
Range: $-\infty < x < \infty$
Mean: m_x
Variance: σ_x^2

2.6.3   *Exponential distribution*

The pdf of an exponential distribution is

$$f(x) = \begin{cases} (1/b)\exp(-x/b), & 0 \le x < \infty, \quad b > 0 \\ 0, & \text{elsewhere} \end{cases} \tag{2.57}$$

Algorithm

1. Generate u from a uniform distribution (0,1)
2. $x = -b \ln(u)$
3. Keep x

Book MATLAB function: sspexponentialpdf(b,N)

```
function[x,m,sd]=sspexponentialpdf(b,N)
%[x,m,sd]=sspexponentialpdf(b,N);
for i=1:N
    x(i)=-b*log(rand);
end;
m=mean(x);sd=std(x);
```

Book MATLAB function: sspexponentialpdf1(b,N)

```
function[x,m,sd]=sspexponentialpdf1(b,N)
x=-b*log([rand(1,N)]);
m=mean(x);sd=std(x);
```

Other statistical characteristics of the exponential distribution:
Range: $0 \le x < \infty$
Mean: b
Variance: b^2

The following m-File creates a histogram for a normal rv (see Figure 2.8). With small modification it can be used for any distribution.

Book MATLAB m-File: normal_hist

```
%normal histogram m-File: normal_hist
n=1000;
x=1.5*randn(1,n)+6*ones(1,n);
subplot(1,2,1);plot(x(1,1:200),'k');
xlabel('n');ylabel('x(n)');grid on;
```

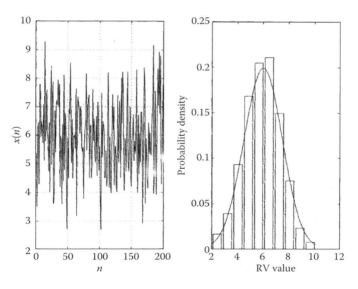

Figure 2.8

```
[m,z]=hist(x,10); %calculates counts in bins
                %and bin coordinates for 20 bins;
w=max(x)/length(z); %calculates bin width;
pb=m/(n*w);     %probability in each bin;
v=linspace(min(x),max(x));%generates 100 values over
                %range of rv x;
y=(1/(2*sqrt(2*pi)))*exp(-((v-6*ones(size(v))).^2)/4.5);
                %normal pdf;
subplot(1,2,2);
colormap([1 1 1]);%creates white bars, for other colors
                % see >>help colormap;
bar(z,pb); %plots histogram;
hold on;plot(v,y,'k')%superimpose plot of normal pdf;
xlabel('RV value');ylabel('Probability density');
```

2.6.4   *Lognormal distribution*

Let the rv x be $N(m_x, \sigma^2)$. Then, the rv $y = \exp(x)$ has the lognormal distribution with the following pdf

$$f(y) = \begin{cases} \left(1/\sqrt{2\pi}\sigma y\right)\exp\left[-((\ln y - m_x)/2\sigma^2)\right], & 0 \le y < \infty \\ 0, & \text{elsewhere} \end{cases} \qquad (2.58)$$

The values of σ_x and m_x must take small values to produce a lognormal-type distribution.

Algorithm

1. Generate z from $N(0,1)$
2. $x = m_x + \sigma_x z$ (x is $N(m_x, \sigma_x^2)$)
3. $y = \exp(x)$
4. Keep y

Book MATLAB function: ssplognormalpdf(m,s,N)

```
function[y]=ssplognormalpdf(m,s,N)
%[y]=ssplognormalpdf(m,s,N);
%m=mean value;s=standard deviation;N=number of samples;
for i=1:N
    r1=rand;
    r2=rand;
    z(i)=sqrt(-2*log(r1))*cos(2*pi*r2);
end;
x=m+s*z;
y=exp(x);
```

Book MATLAB function: ssplognormalpdf1

```
function[y]=ssplognormalpdf1(m,s,N)
n=sqrt(-2*log(rand(1,N))).*cos(2*pi*rand(1,N));
x=m+s*n;
y=exp(x);
```

Other statistical characteristics of the lognormal distribution (s = standard deviation):

Range: $0 \le x < \infty$
Mean: $\exp(m_x)\exp(s^2/2)$
Variance: $\exp(2m_x)\exp(s^2)(\exp(s^2)-1)$

2.6.5 Chi-square distribution

If $x_1, x_2, \ldots, x_N$ is a random sample of size r from a distribution that is $N(0,1)$, then

$$y = \sum_{n=1}^{r} x_n^2 \qquad (2.59)$$

has a chi-square distribution with **r degrees of freedom**.
The pdf of the chi-square distribution is

$$f(x) = \frac{1}{\Gamma(r/2)2^{r/2}} x^{(r/2)-1} e^{-x/2} \qquad (2.60)$$

where r is the number of degrees of freedom and $\Gamma()$ is the gamma function. MATLAB uses the function **gamma()** for evaluating the gamma function.

Book MATLAB function for chi-square distribution: sspchisquared pdf(n,df)

```
function[y]=sspchisquaredpdf(n,df)
%Book MATLAB function:y=sspchisquaredpdf(n,df);
%n=number of the chi-distributed rv y;df=degrees
%of freedom, MUST BE EVEN number;
for m=1:n
    for i=1:df/2
        u(i)=rand;
    end;
    y(m)=-2*log(prod(u));
end;
```

We can also use a Monte Carlo approach to find rvs that are chi-squared distributed. The following program does this.

Book MATLAB function: sspmontecarlochisquaredpdf(r,N)

```
function[y]=sspmontecarlochisquaredpdf(r,N)
%N=number of chi-squared distributed rv's y;
%r=degrees of freedom, MUST BE EVEN number;
for n=1:N
    y(n)=sum(randn(1,r).^2);
end;
```

Other statistical characteristics of the chi-square distribution:
    Range: $0 \leq x < \infty$
    Mean: r
    Variance: $2r$

2.6.6 *Student's* t *distribution*

The pdf of the Student's t distribution is

$$f(x) = \frac{\Gamma[(r+1)/2]}{\sqrt{\pi r}\,\Gamma(r/2)} \frac{1}{(1+(x^2/r))^{((r+1)/2)}}, \quad -\infty < x < \infty \tag{2.61}$$

where $r = N - 1$ is the number of degrees of freedom and N is the number of terms in the sequence $\{x(n)\}$.

If z has a standard normal distribution, $N(0,1)$, and y has a chi-square distribution with r degrees of freedom, then

$$x = \frac{z}{\sqrt{y/r}} \tag{2.62}$$

has Student's distribution with r degrees of freedom. To generate x, we first generate z, as described above for the Gaussian distribution; then, we generate y as described above for the chi-square distribution and apply Equation 2.62.

Book MATLAB function: ssptdistributionpdf(r,N)

```
function[t]=ssptdistributionpdf(r,N)
%r=degrees of freedom,N=number of iid variables;
z=randn(1,N);
y=sspchisquaredpdf(N,r);
t=z./sqrt(y/r);
```

Other statistical properties of the t distribution:
  Mean: 0
  Variance: $c/(c-2)$, $c>2$

2.6.7   F *distribution*

If y_1 is a chi-square rv with r_1 degrees of freedom and y_2 is chi-square rv with r_2 degrees of freedom and both rvs are independent, then the rv

$$x = \frac{y_1/r_1}{y_2/r_2}, \quad 0 \le x < \infty \tag{2.63}$$

is distributed as an F distribution. To create an F variate, we first generate two chi-square variates and apply Equation 2.63.

The pdf of the F distribution is

$$f(x) = \frac{\Gamma[(r_1+r_2)/2]}{\Gamma(r_1/2)\Gamma(r_2/2)} \left(\frac{r_1}{r_2}\right)^{r_1/2} \frac{x^{(r_1/2)-1}}{(1+(r_1/r_2)x)^{((r_1+r_2)/2)}}, \quad 0 \le x < \infty \tag{2.64}$$

where $\Gamma(\cdot)$ is the gamma function.

2.6.8   *Rayleigh probability density function*

A radio wave field, which arrives at a receiving point (antenna) after been scattered from a number of scattering points (trees, buildings, etc.), is given by the relation:

$$S = Re^{j\theta} = \sum_{k=1}^{n} A_k e^{j\Phi_k} \tag{2.65}$$

where
  Φ_k is an uniformly distributed phase
  A_ks are identically distributed

Resolving S into its real and imaginary components, we have

$$X = \text{Re}\{S\} = R\cos\theta = \sum_{k=1}^{n} A_k \cos\Phi_k = \sum_{k=1}^{n} X_k \tag{2.66}$$

$$Y = \text{Im}\{S\} = R\sin\theta = \sum_{k=1}^{n} A_k \sin\Phi_k = \sum_{k=1}^{n} Y_k \tag{2.67}$$

In general, n is large and, therefore, the central limit theorem dictates that both X and Y are normally distributed. If, in addition, we assume that A_ks and Φ_ks are independent, then the means of X and Y are

$$E\{X\} = \sum_{k=1}^{n} E\{A_k\}E\{\cos\Phi_k\} = \sum_{k=1}^{n} E\{A_k\}\frac{1}{2\pi}\int_{c}^{c+2\pi} \cos\phi_k \, d\phi_k = 0 \tag{2.68}$$

and, similarly,

$$E\{X\} = \sum_{k=1}^{n} E\{A_k\}E\{\sin\Phi_k\} = \sum_{k=1}^{n} E\{A_k\}\frac{1}{2\pi}\int_{c}^{c+2\pi} \sin\phi_k \, d\phi_k = 0 \tag{2.69}$$

Since $E\{X\} = 0$ and $E\{Y\} = 0$, the variances are

$$E\{X^2\} = \sum_{k=1}^{n} E\{A_k^2\}E\{\cos^2\Phi_k\} = \frac{1}{2}nE\{A_k^2\} \tag{2.70}$$

$$E\{Y^2\} = \sum_{k=1}^{n} E\{A_k^2\}E\{\sin^2\Phi_k\} = \frac{1}{2}nE\{A_k^2\} \tag{2.71}$$

Setting

$$\frac{1}{2}nE\{A_k^2\} = \sigma^2 \tag{2.72}$$

we obtain the pdfs of X and Y as

$$f(x) = \frac{1}{\sigma\sqrt{2\pi}}e^{-x^2/2\sigma^2}$$

$$f(y) = \frac{1}{\sigma\sqrt{2\pi}}e^{-y^2/2\sigma^2} \tag{2.73}$$

To find the pdf $f(x, y)$, we must first find if X and Y are correlated. Therefore, we investigate the ensemble average

$$E\{XY\} = \sum_{k}^{n} \sum_{l}^{n} E\{A_k A_l\} E\{\cos \Phi_k \sin \Phi_l\} = 0 \qquad (2.74)$$

since $E\{\cos \Phi_k \sin \Phi_l\} = 0$ for all k and l. Therefore, X and Y are uncorrelated and since they are normal they are also independent. Hence, the combined pdf is

$$f(x,y) = f(x)f(y) = \frac{1}{\sigma\sqrt{2\pi}} e^{-x^2/2\sigma^2} \frac{1}{\sigma\sqrt{2\pi}} e^{-y^2/2\sigma^2}$$

$$= \frac{1}{2\pi\sigma^2} e^{-(x^2+y^2)/2\sigma^2} \qquad (2.75)$$

Transformation of PDFs

Consider the functions:

$$U = U(X,Y), \qquad V = V(X,Y) \qquad (2.76)$$

and their inverse functions:

$$X = X(U,V), \qquad Y = Y(U,V) \qquad (2.77)$$

If the rv X is near the value x and the rv Y is near the value y, U and V must be near $u = u(x,y)$ and $v = v(x,y)$ given by the functions (Equation 2.76). Hence,

$$p\{x < X \le x+dx, y < Y \le y+dy\} = f_{XY}(x,y)dx\,dy = p\{u(x,y) < U \le u(x,y)$$
$$+ du(x,y), v(x,y) < V \le v(x,y) + dv(x,y)\}$$
$$= f_{UV}(u,v)du\,dv \qquad (2.78)$$

Substituting the inverse functions from Equation 2.77 into above equation and dividing by $du\,dv$ both sides, we obtain the required pdf to be equal to

$$f_{UV}(u,v) = f_{XY}[x(u,v), y(u,v)] \left| \frac{dx\,dy}{du\,dv} \right| \qquad (2.79)$$

where the absolute value is necessary to take into consideration both the nondecreasing and nonincreasing pdfs. The quantity shown with the absolute value symbol is known as the **Jacobian** of the transformation.

The equivalent one-dimensional transformation is given by

$$f_Y(y) = f_X[f^{-1}(y)] \left| \frac{dx}{dy} \right|, \quad y = f(x), \quad x = f^{-1}(y) = \text{inverse function} \qquad (2.80)$$

To transform Equation 2.75 to polar coordinates, we use the following equations (correspond to Equation 2.76):

$$R(X,Y) = \sqrt{X^2 + Y^2}, \quad \Theta(X,Y) = \tan^{-1}(Y/X) \qquad (2.81)$$

with inverse functions:

$$X(R,\Theta) = R\cos\Theta, \quad Y(R,\Theta) = R\sin\Theta \qquad (2.82)$$

The Jacobian of this transformation is

$$\frac{\partial(x,y)}{\partial(r,\theta)} = \begin{vmatrix} (\partial x/\partial r) & (\partial x/\partial \theta) \\ (\partial y/\partial r) & (\partial y/\partial \theta) \end{vmatrix} = \begin{vmatrix} \cos\theta & -r\sin\theta \\ \sin\theta & r\cos\theta \end{vmatrix} = r \qquad (2.83)$$

Substituting Equations 2.82 and 2.83 into Equation 2.79, we obtain

$$f(x,y) = \frac{r}{2\pi\sigma^2} e^{-r^2/2\sigma^2}, \quad 0 \le \theta \le 2\pi, \quad 0 \le r < \infty \qquad (2.84)$$

If we set $\alpha = 2\sigma^2$ in the above equation, we obtain the general form of the Rayleigh pdf, which is given by

$$f(r,\theta) = \frac{r}{\pi\alpha} e^{-r^2/\alpha} \qquad (2.85)$$

The distribution of θ is, as expected, uniform:

$$f(\theta) = \int_{r=0}^{\infty} f(r,\theta) dr = \frac{1}{\pi\alpha} \int_{r=0}^{\infty} r e^{-r^2/\alpha} dr = \frac{1}{\pi} \left[-\frac{1}{2} \int_{r=0}^{\infty} \frac{d}{dr} e^{-r^2/\alpha} dr \right]$$

$$= -\frac{1}{2\pi} \left[e^{-r^2/\alpha} \Big|_{r=0}^{\infty} \right] = -\frac{1}{2\pi} [0-1] = \frac{1}{2\pi}, \quad 0 \le \theta \le 2\pi \qquad (2.86)$$

The distribution for the rv R is

$$f(r) = \int_{0}^{2\pi} \frac{r}{\pi\alpha} e^{-r^2/\alpha} d\theta = \frac{2r}{\alpha} e^{-r^2/\alpha}, \quad r \ge 0 \qquad (2.87)$$

i.e., the Rayleigh distribution.

2.7   Estimators

Most, if not all the times, of our measurements produce a **sample** of N data values (a sequence) $\{x(n)\}$. We perform mathematical operations to determine some of the statistical properties, e.g., mean, variance, etc. These values are built around the notion of **point estimate**. The whole process is known as **estimation**. Since each realization will produce different values of these statistical quantities, we are confronted with a **parameter space**.

Consider a family of pdfs $\{f(x;\theta);\ \theta \in \Theta\}$. In practice, we have only one member of the family of sequences and, therefore, we need a point estimate of θ. What we do next, is to find a **statistic** (a function):

$$\hat{\theta} = u(x_1, x_2, \ldots, x_N) \tag{2.88}$$

where x_is are the values of the corresponding rvs. Then, the number $\hat{\theta}$ will be a good point estimate.

Quality of estimators

To qualify $\hat{\theta} = u(x_1, x_2, \ldots, x_N)$ as a good point estimator of θ, there must be a high probability that the estimate value $\hat{\theta}$ is close to the population one, θ. One of the conditions is that the expectation of the statistic should be equal to the population parameter θ, hence we must have

$$E\{\hat{\theta}\} = \theta \tag{2.89}$$

Under this circumstance, the estimator is called an **unbiased** estimator. The **biased** estimator defines the relation:

$$E\{\hat{\theta}\} - \theta \neq 0 \quad \text{or} \quad B(\hat{\theta}) = E\{\hat{\theta}\} - \theta \tag{2.90}$$

where $B(\hat{\theta})$ is the bias.

Example 2.11

Let the finite sequence $\{x(n)\}$ be made of N iid rvs. Show that the sample mean is an unbiased estimator.

Solution

We proceed as follows:

$$E\{\hat{m}_x\} = E\{u(x_1, \ldots, x_N)\} = E\left\{\frac{1}{N}\sum_{n=1}^{N} x(n)\right\} = \frac{1}{N}\sum_{n=1}^{N} E\{x(n)\}$$

$$= \frac{1}{N}\sum_{n=1}^{N} m_x = \frac{Nm_x}{N} = m_x$$

Therefore, the sample mean is an unbiased estimator. We observe that the sample mean is a *statistic* for the *mean* parameter. ■

Unbiased minimum variance

For a given positive integer N, $y = u(x_1, x_2, ..., x_N)$ is called an **unbiased minimum variance** estimator of the parameter θ if y is unbiased, i.e., $E\{y\} = \theta$, and if the variance of y is less than or equal to the variance of every other unbiased estimator of θ.

Example 2.12

Let the signal be

$$x(n) = A + v(n)$$

where $v(n)$ is a zero-mean noise. Based on the data set $\{x(0)\, x(1) \cdots x(N-1)\}$, we would like to estimate A. Since the average value of $v(n)$ is 0, it is natural to estimate A by the time average value (sample mean) of the data, or

$$\hat{A} = \frac{1}{N} \sum_{n=0}^{N-1} x(n)$$

We have already shown in the previous example that $E\{\hat{A}\} = A$, which indicates that the accepted statistic (sample mean) is an unbiased estimator. Next, we proceed to find the variance of the sample mean. Hence,

$$\mathrm{var}\{\hat{A}\} \triangleq E\{\hat{A} - A\} = E\left\{ \left(\frac{1}{N} \sum_{n=0}^{N-1} x(n) - A \right)^2 \right\}$$

$$= E\left\{ \frac{1}{N^2} \sum_{n=0}^{N-1} (x(n) - A)^2 + \frac{1}{N^2} \sum_{n=0}^{N-1} \sum_{m=0}^{N-1} (x(n) - A)(x(m) - A) \right\}$$

$$= \frac{1}{N^2} \sum_{n=0}^{N-1} E\{x(n) - A)^2\} = \frac{N\sigma^2}{N^2} = \frac{\sigma}{N}$$

where in the double summation $n \neq m$. Because $(x(n) - A)(x(m) - A)$ are independent $E\{(x(n) - A)(x(m) - A)\} = (A - A)(A - A) = 0$. This result indicates that as the number of data increases, the mean square value difference from the population mean decreases and the estimate becomes more accurate. ▨

If the bias and the variance of an estimator tend to zero as the sample size N becomes large. This means that the sampling distribution will tend to be centered about A (in the above examples) and the precision of the estimator increases without limit. An estimator possessing this property is called a **consistent estimator.**

2.8   Confidence intervals

Confidence intervals of the sample mean

We are, next, interested in finding confidence intervals for the sample mean estimator. We know from theory of statistics that the sample mean and variance of an iid sequence $\{x(n)\}$ with normal distribution, $N(m, \sigma^2)$, have the following properties (see Problem 2.8.1):

$$\hat{m}_x \text{ is } N(m, \sigma^2/N) \tag{2.91a}$$

$$N\hat{\sigma}^2/\sigma^2 \text{ is } \chi^2(N-1), \qquad \hat{\sigma}^2 = \sum_{n=1}^{N} \frac{(\hat{m}_x - m_x)^2}{N} \tag{2.91b}$$

$$\hat{m}_x \text{ and } \hat{\sigma}^2 \text{ are stochastically independent} \tag{2.91c}$$

Therefore, we have that $\sqrt{N}(\hat{m}_x - m_x)/\sigma$ is $N(0, 1)$, and $N\hat{\sigma}^2/\sigma^2$ is $\chi^2(N-1)$ and both are statistically independent. The above t distribution was defined as the ratio of such two rvs. Based on the t distribution and the above discussion, we know that

$$t = \frac{\left[\sqrt{N}(\hat{m}_x - m_x)\right]/\sigma}{\sqrt{N\hat{\sigma}^2/[\sigma^2(N-1)]}} = \frac{\hat{m}_x - m_x}{\hat{\sigma}/\sqrt{N-1}} \tag{2.92}$$

has a t distribution with $N - 1$ degrees of freedom, for all $\hat{\sigma}^2 > 0$. Table 2.1 gives us some feeling of the critical region that we may want to select. For example, we want to find out the probability that the mean value falls within a range, say, 0.95% or 95% probable. It is instructive to see Figure 2.9 for some range and critical region definitions.

Since the area under a pdf is 1, we have the following relationships:

$$\int_{-\infty}^{-t_c} f(t)dt = 1 - \int_{-\infty}^{t_c} f(t)dt$$

$$\text{Confidence region}(-t_c < t < t_c) = 1 - 2\left(\int_{t_c}^{\infty} f(t)dt\right) \tag{2.93}$$

Table 2.1 The t Distribution (see Section 2.6)

Degrees of Freedom	$p\{T \le t\}$		
	0.90	0.95	0.975
r	t_c	t_c	t_c
2	1.886	2.920	4.303
5	1.476	2.015	2.571
8	1.397	1.860	2.306
12	1.356	1.782	2.179
15	1.341	1.753	2.131
20	1.325	1.725	2.086
25	1.316	1.708	2.060
30	1.310	1.697	2.042

Next, we would like to find the probability that the population mean will be within the confidence region. From Equation 2.92, we obtain

$$p\left(-t_c < \frac{\hat{m}_x - m_x}{\hat{\sigma}/\sqrt{N-1}} < t_c\right) \quad \text{or} \quad p\left(\hat{m}_x - \frac{t_c\,\hat{\sigma}}{\sqrt{N-1}} < m_x < \hat{m}_x + \frac{t_c\,\hat{\sigma}}{\sqrt{N-1}}\right) \qquad (2.94)$$

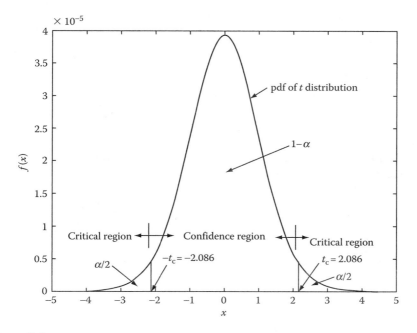

Figure 2.9

Using MATLAB and assuming that the statistics of 10,000 terms sequence gives the population mean, we can obtained a 40 term subsequence. Next, if we want the area of the critical region to be about $\alpha = 0.05$, and selecting a critical time $t_c = 2.0$, we find that the area of the critical region (using MATLAB for the integration) is

$$\alpha = 2\int_2^\infty f(t)dt = 0.0525$$

This value is very close to the desired one. If the critical value was not guessed correctly, we can try repeating the integration for other values until we get a good approximation. Since in the command window we have to change only t_c, we can perform each new integration within a few second. Using this approach, we do not need tables. Therefore, the probability to have the population mean between -2 and 2 is $1 - 0.0525 = 0.9475\%$ or 94.75% that its value will be within that range. Inserting the values found in MATLAB for the sequence of 40 terms, we obtain

$$p\left(\hat{m}_x - \frac{t_c\,\hat{\sigma}}{\sqrt{N-1}} < m_x < \hat{m}_x + \frac{t_c\,\hat{\sigma}}{\sqrt{N-1}}\right)$$

$$= p\left(0.5148 - \frac{2\times 0.0914}{\sqrt{40-1}} < m_x < 0.5148 + \frac{2\times 0.0914}{\sqrt{40-1}}\right)$$

$$= p(0.4855 < m_x < 0.5441) = 0.9475 \tag{2.95}$$

The population mean was found to be equal to $m_x = 0.5052$.

Confidence intervals for the variance

For a realization, $\{x(n)\}$, with elements $N(m_x, \sigma^2)$, it is desired to find the confidence interval for σ^2. It is assumed, as is always in practice, that the population mean is not known. The identification of the region can be accomplished using the estimate of the population variance σ^2 and $N\hat{\sigma}^2/\sigma^2$ which is χ^2 $(N-1)$ distributed. For a particular sequence with length N, we must find positive constants a and b such that the probabilities

$$p\left(a < \frac{N\hat{\sigma}^2}{\sigma^2} < b\right), \quad a < b \quad \text{or} \quad p\left(\frac{N\hat{\sigma}^2}{b} < \sigma^2 < \frac{N\hat{\sigma}^2}{a}\right) \tag{2.96}$$

are equal to some desired value, e.g., 0.95. Table 2.2 gives some values involving the χ^2 distribution.

Example 2.13

Find the x_cs so that the population variance has a 95% probability to fall within the confidence region. The

Table 2.2 Chi-Square Distribution (see Section 2.6)

Degrees of Frequency	$p\{T \le t\}$			
	0.050	0.95	0.975	0.99
r	x_c	x_c	x_c	x_c
5	1.15	11.1	12.8	15.1
8	2.73	15.5	17.5	20.1
10	3.94	18.3	20.5	23.2
12	5.23	21.0	23.3	26.2
15	7.26	25.0	27.5	30.6
20	10.9	31.4	34.2	37.6
25	14.6	37.7	40.6	44.3
30	18.5	43.8	47.0	50.9

sequence is x = {−0.4326 −1.6656 0.1253 0.2877 −1.1465 1.1909 1.1892 −0.0376 0.3273 0.1746}.

Solution

For this case $N = 9$, $\hat{\sigma}^2 = 0.8162$, and if we take $a = 17$ and $b = 2$, we obtain (using MATLAB) that the probability $p(2 < X < 17)$ is 0.9428, which is very close to the desired one. Hence, Equation 2.96 gives

$$p\left(\frac{9 \times 0.8162}{17} < \sigma^2 < \frac{9 \times 0.8162}{2}\right)$$

$$= p(0.4321 < \sigma^2 < 3.6729) = 0.9428 \qquad (2.97)$$

The population variance for this case is equal to 1.0023. ■

Problems

2.1.1 Create the following two sequences using MATLAB: (a) x = randn(1,64) and (b) y = conv(x,[1 0.8 0.2 0.1]). Plot the following: (1) plot($x(n)$, $x(n + 1)$,'x'); (2) plot($x(n)$, $x(n + 3)$,'x'); (3) plot($y(n)$, $y(n + 1)$,'x'); (4) plot($y(n)$, $y(n + 3)$, 'x'). Observe the similarities and differences and indicate the correlated and uncorrelated situations.

2.1.2 Plot the cumulative pdf of the central normal (Gaussian) pdf: $f(x) = \exp(-0.5x^2)/\sqrt{2\pi}$, $\sigma = 1$. Plotted from $x = -3$ to 3.

2.2.1 Using the continuous case formulation find the sample mean value formula.

2.2.2 Show that the sample mean is equal to the population mean.

2.2.3 The autocorrelation of a continuous rv is defined as follows:

$$R_x(t+\tau,t) = E\{X(t+\tau)X(t)\}$$

Using the above equation find the mean value and the autocorrelation function of the random function $X(t) = A\sin(\omega t + \varphi)$ where $A, \omega > 0$ are constants, and φ is a uniform rv over the range $0 < \varphi < 2\pi$. Find the mean and autocorrelation of $X(t)$.

2.3.1 Verify Equation 2.18.

2.3.2 Find the autocorrelation function $x(n) = A\cos(n\omega + \theta)$, where A and ω are constants and θ is uniformly distributed over the interval $-\pi$ to π.

2.3.3 Prove the following properties of a WSS process: (a) $m(n) = m_n = $ constant; (b) $r_x(-k) = r_x(k)$; (c) $r_x(m,n) = r_x(m-n)$; (d) $r_x(0) > r_x(k)$.

2.3.4 Based on the results of Problem 2.3.2, create a 2×2 autocorrelation matrix.

2.3.5 If the sequence $\{x(n)\}$ is characterized by iid rvs find the mean and the variance of the signal $y(n) = ax(n) + b$, where a and b are constants.

2.3.6 Verify Equation 2.24.

2.3.7 Verify the covariance of the RW expression.

2.3.8 Repeat Example 2.6 for the following sequences: (a) $x = $ randn(1,128); (b) $x = $ rand(1,128); (c) $x = $ rand(1,128) $- 0.5$; (d) $x = $ conv(randn(1,125), [1 0.8 0.2 0.1]). Repeat the above steps using windows, e.g., $xw = x^*$. window(@hamming,length(x)).

2.3.9 Given the first-order MA process, $y(n) = b(0)x(n) + b(1)x(n-1)$, find $r_y(0)$ and $r_y(1)$.

2.4.1 Verify Equation 2.42.

2.5.1 Find the output autocorrelation function of a system if the input is a WSS process. The input, output, and the impulse response of the system are, respectively, $x(n)$, $y(n)$, and $h(n)$.

2.5.2 If the rv x is $N(m_x,\sigma^2)$ with variance greater than zero, then the rv $w = (x - m_x)/\sigma$ is $N(0,1)$.

2.6.1 (a) Plot the pdfs for the t distribution and the F distribution. Use different values of degrees of freedom. (b) Find the $p\{x \le 2.5\}$ for a t distribution and 16 degrees of freedom.

2.7.1 The following random sequence is given having iid elements with Gaussian distribution: $x = \{$1.4279 3.3335 1.3186 1.0087 $-$0.4647 1.7888 $-$1.4724 2.6286 4.4471 $-$0.1836$\}$. Find the region of confidence so that there is 95% probability that the population mean will fall in this region.

2.8.1 Verify Equation 2.91a.

Solutions, suggestions, and hints

2.1.1 The rvs $y(n)$ are the output of a low-pass filter.

2.1.2 Use the following MATLAB m-file:

```
% file: prob2_1_2
for x=1:120
    y=1:x;
    F(x)=0.05*sum(exp(-0.5*(y*0.05-3).^2))/sqrt(2*pi);
end;
m=-3:0.05:3-0.05;
f=exp(-m.^2/2)/sqrt(2*pi);
plot(m,F,m,f)
```

2.2.1 In this development, we will drop the time n and, hence, it is assumed that we deal with a rv x at time n. Let us divide the real line x into intervals of length Δx with boundary points x_i. We further assume that we want to find the mean value of the function $g(x)$. Hence, Equation 2.7 becomes

$$E\{g(x)\} = \int_{-\infty}^{\infty} g(x)f(x)dx \cong \sum_{i=-\infty}^{\infty} g(x_i)f(x_i)\Delta x$$

To make the transformation from the integral to summation, we made the following important assumption: all the sample values of x have the same pdf, which means that they are identically distributed. Next, we see that $f(x_i)\Delta x \cong p(x_i \leq x < x_i + \Delta x)$, where x_i is the value at time n (we have suppressed the time here). Using the relative frequency definition of the probability, we obtain (see Equation 2.9):

$$E\{g(x)\} \cong \sum_{i=-\infty}^{\infty} g(x_i)p(x_i \leq x < x_i + \Delta x) = \sum_{i=-\infty}^{\infty} g(x_i)\frac{N_i}{N}$$

where N_i is the number of measurements in the interval: $x_i \leq x < x_i + \Delta x$. Because $g(x_i)N_i$ approximates the sum of values of $g(x)$ for points within the interval i, then $E\{g(x)\} \cong (1/N)\sum_{i=1}^{N} g(x_i)$. If $g(x) = x$, then we obtain $E\{x\} \cong \hat{m} = (1/N)\sum_{i=1}^{N} x_i$.

2.2.2 $E\{\hat{m}\} = E\left\{\dfrac{1}{N}\sum_{k=1}^{N} x(k)\right\} = \dfrac{1}{N}\sum_{k=1}^{N} E\{x(k)\} = \dfrac{1}{N}\sum_{k=1}^{N} m = \dfrac{Nm}{N} = m.$

2.2.3 Mean:

$$E\{X(t)\} = E\{A\sin(\omega t + \varphi)\} = \int_{-\infty}^{\infty} A\sin(\omega t + \varphi)f(\varphi)d\varphi = \frac{A}{2\pi}\int_{-\pi}^{\pi} \sin(\omega t + \varphi)d\varphi$$

$$= \frac{A}{2\pi}[-\cos(\omega t + \varphi)]\big|_{\varphi=-\pi}^{\pi} = 0$$

Autocorrelation: $\sin a \sin b = (1/2)[\cos(a - b) - \cos(a + b)]$
Therefore, we have

$$R_x(t + \tau, t) = E\left\{ A^2 \underbrace{\sin(\omega t + \omega \tau + \varphi)}_{a} \underbrace{\sin(\omega t + \varphi)}_{b} \right\}$$

$$= \frac{A^2}{2} E\{\cos(\omega \tau)\} - E\{\cos(2\omega t + \omega \tau + 2\varphi)\}$$

$$= \frac{A^2}{2}\left[\cos(\omega \tau) - \int_{-\infty}^{\infty} \cos(2\omega t + \omega \tau + 2\varphi) f(\varphi) d\varphi \right]$$

$$= \frac{A^2}{2}\left[\cos(\omega \tau) - \int_{-\pi}^{\pi} \cos(2\omega t + \omega \tau + 2\varphi) \frac{1}{2\pi} d\varphi \right]$$

$$= \frac{A^2}{2} \cos(\omega \tau) = R_x(\tau)$$

Note that the mean value is a constant and the correlation function is a function of the time difference, τ.

2.3.1 (a) $E\{x(n + k)\} = E\{x(m + k)\}$ implies that the mean value must be constant independent of time.

(b) $r_x(m, n) = \iint x(m)x(n) f(x(m + k)x(n + k)) dx(m) dx(n) = r_x(m + k, n + k)$
$= r_x(m - n, 0) = r_x(m - n)$.

(c) $r_x(k) = E\{x(n + k)x(n)\} = E\{x(n)x(n + k)\} = r_x(n - n - k) = r_x(-k)$.

(d) $E\{[x(n + k) - x(n)]^2\} = r_x(0) - 2r_x(k) + r_x(0) \geq 0$ (because of the square)
or (because of the square) or $r_x(0) \geq r_x(k)$.

2.3.2 $m_n = E\{x(n)\} = E\{A \sin(n\omega + \theta)\} = A \int_{-\pi}^{\pi} \sin(n\omega + \theta)(1/2\pi)d\theta = 0$.

$r_x(m, n) = E\{A^2 \sin(m\omega + \theta)\sin(n\omega + \theta)\}$
$= (1/2)A^2 E\{\cos[(m - n)\omega] - (1/2)\cos[(m + n)\omega + 2\theta]\}$
$= (1/2)A^2 \cos[(m - n)\omega]$.

Because the ensemble average of the sine function with respect to θ is 0 (constant) and the autocorrelation is a function of the lag factor, the signal is WSS process.

2.3.3 (a) $E\{x(n + q)\} = E\{x(m + q)\}$ implies that the mean must be a constant;
(b) $r_x(m, n) = \iint x(m)x(n) f(x(m + k), x(n + k)) dx(m) dx(n)$
$= r_x(m + k, n + k) = r_x(m - n, 0) r_x(m - n);$

(c) $r_x(k) = E\{x(n+k)x(n)\} = E\{x(n)x(n+k)\} = r_x(n-n-k) = r_x(-k);$

(d) $E\{[x(n+k)-x(n)]^2\} = r_x(0) - 2r_x(k) + r_x(0) \geq 0 \quad \text{or} \quad r_x(0) \geq r_x(k).$

2.3.4 $\mathbf{R}_x = \dfrac{A^2}{2}\begin{bmatrix} 1 & \sin(\omega) \\ \sin(\omega) & 1 \end{bmatrix}.$

2.3.5 $E\{y(n)\} = m_y = E\{ax(n)+b\} = aE\{x(n)\} + E\{b\} = am_x + b;$

$$\begin{aligned} E\{y(n)y(n)\} &= E\{[ax(n)+b-m_y][ax(n)+b-m_y]\} \\ &= E\{a^2x^2(n) + b^2 + a^2m_x^2 + b^2 + 2abx(n) - 2a^2m_xx(n) \\ &\quad - 2abx(n) - 2abm_x - 2b^2 + 2abm_x\} \\ &= a^2[E\{x^2(n)\} - m_x^2] = a^2\sigma_x^2 \end{aligned}$$

2.3.6 (a) $E\{y(n)\} = b_0E\{x(n)\} + b_1E\{x(n-1)\} + \cdots + E\{x(n-q)\} = 0$ since it is assumed that the mean value of the rvs $x(n)$s is 0.

(b) $$\begin{aligned} \operatorname{var}(y(n)) &= E\{y(n)y(n)\} = E\{(b_0x(n) + b_1x(n-1) + \cdots + b_qx(n-q))^2\} \\ &= b_0^2E\{x^2(n)\} + b_1^2E\{x^2(n-1)\} + \cdots + 2\sum_{i\neq j} b_ib_jE\{x(n-i)x(n-j)\} \\ &= b_0^2\sigma_x^2 + b_1^2\sigma_x^2 + \cdots + 2\sum_{i\neq j} b_ib_j0 = \sigma_x^2\sum_{i=0}^{q} b_i^2 \end{aligned}$$

since the rvs are independent with mean zero.

2.3.7 $$c(y(n), y(n)) = E\left\{\sum_{i=1}^{n} x(i)\sum_{k=1}^{n} x(k)\right\} = \sum_{i=1}^{n}\sum_{k=1}^{n} E\{x(i)x(k)\}$$

$$= \begin{cases} \sum_{i=1}^{n}\sum_{k=1}^{n} E\{x(i)\}E\{x(k)\} = 0, & i \neq k \\ \sum_{i=1}^{n} E\{x^2(i)\} = \sum_{i=1}^{n} \sigma_x^2 = n\sigma_x^2, & i = k \end{cases}$$

2.3.9 $$\begin{aligned} r_y(0) &= E\{[b(0)x(n) + b(1)x(n-1)][b(0)x(n) + b(1)x(n-1)]\} \\ &= b(0)^2E\{x(n)^2\} + b(1)^2E\{x(n-1)^2\} + 2b(0)b(1)E\{x(n)x(n-1)\} \\ &= b(0)^2\sigma_x^2 + b(1)^2\sigma_x^2 \end{aligned}$$

since WN with zero mean value has the property $E\{x(n)\,x(n-1)\} = E\{x(n)$ $E\{x(n-1)\} = 0$. Similarly, we obtain $r_y(1) = E\{[b(0)x(n) + b(1)x(n-1)][b(0)$ $x(n-1) + b(1)x(n-2)]\} = b(0)b(1)\sigma_x^2$.

2.4.1 $S_x(e^{j\omega}) = \sum_{k=-\infty}^{\infty} r_x(k)e^{-j\omega k} = \sum_{k=0}^{\infty} a^k e^{-j\omega k} + \sum_{k=0}^{\infty} a^k e^{j\omega k} - 1$

where for the negative values of k we set $k = -k$ in the expression. Since to the negative summation we introduced an extra $k = 0$ that gives 1 we must subtract 1 from the expression. Applying the geometric series formula for infinite length, we obtain

$$S_x(e^{j\omega}) = \frac{1}{1 - ae^{-j\omega}} + \frac{1}{1 - ae^{j\omega}} - 1 = \frac{1 - a^2}{1 - 2a\cos\omega + a^2}$$

2.5.1 $y(n) = x(n) * h(n) = \sum_{k=-\infty}^{\infty} h(k)x(n-k)$ (1); $E\{y(n)\} = \sum_{k=-\infty}^{\infty} h(k)E\{x(n-k)\}$

$= m_x \sum_{k=-\infty}^{\infty} h(k)$ (2). The ensemble operates only on rvs.

Next, we shift (1) by k, multiply by $x(n)$, and then take the ensemble average of both sides of the equation. Hence, we find

$$E\{y(n+k)x(n)\} = r_{yx}(n+k,n) = E\left\{\sum_{l=-\infty}^{\infty} h(l)x(n+k-l)x(n)\right\}$$

$$= \sum_{l=-\infty}^{\infty} h(l)E\{x(n+k-l)x(n)\}$$

$$= \sum_{l=-\infty}^{\infty} h(l)r_x(k-l) \quad (3) \text{ (remember that } x(n) \text{ is stationary)}$$

Since in (3) the summation is a function only of k, it indicates that the cross-correlation function is also only a function of k. Hence, we write $r_{yx}(k) = r_x(k) * h(k)$ (4). The autocorrelation of $y(n)$ is found from (1) by shifting the time by k, $y(n + k)$, next multiply the new expression by $y(n)$ and then take the ensemble average. Therefore, we find

$$r_y(n+k,n) = E\{y(n+k)y(n)\} = E\left\{y(n+k)\sum_{l=-\infty}^{\infty} x(l)h(n-l)\right\}$$

$$= \sum_{l=-\infty}^{\infty} E\{y(n+k)x(l)\}h(n-l) = \sum_{l=-\infty}^{\infty} h(n-l)r_{yx}(n+k-l) \qquad (5)$$

Here, the input is a WSS process and, therefore, the output is also. Next, we set $m = n - 1$ in (5) to obtain $r_y(k) = \sum_{m=-\infty}^{\infty} h(m)r_{yx}(m+k) = r_{yx}(k) * h(-k)$ (6). Remember that the convolution of two sequences, one been reversed, is equal to the correlation between the original sequences. Combine (6) and (4) to obtain

$$r_y(k) = r_x(k) * h(k) * h(-k) = \sum_{l=-\infty}^{\infty}\sum_{m=-\infty}^{\infty} h(l)r_x(m-l+k)h(m) = r_x(k) * r_h(k)$$

2.5.2 $F(w) = \text{cdf} = \Pr\left\{\dfrac{x - m_x}{\sigma} \leq w\right\} = \Pr\{x \leq w\sigma + m_x\}$

$$= \int_{-\infty}^{w\sigma + m_x} \frac{1}{\sigma\sqrt{2\pi}} \exp\left[-\frac{(x - m_x)}{2\sigma^2}\right] dx$$

If we change the variable of integration by setting $y = (x - m_x)/\sigma$, then $F(w) = \int_{-\infty}^{w} \left(1/\sqrt{2\pi}\right) e^{-y^2/2} \, dy$. But $f(w) = dF(w)/dw$ and, hence, $f(w)$ is equal to the integrand which is the desired solution.

2.7.1 Hint: Try at first critical $t = 2.228$.

2.8.1 The mean value has already been proved in Example 2.11. For the variance (the variables are independent) we have

$$\text{var}\left(\sum_{i=1}^{N} a_i x_i\right) = \sum_{i=1}^{N}\sum_{j=1}^{N} a_i a_j \, \text{cov}(x_i, x_j) = \sum_{i=1}^{N} a_i^2 \, \text{var}(x_i) + 2\sum_{i=1}^{N}\sum_{j>i}^{N} a_i a_j \, \text{cov}(x_i, x_j)$$

$$= \sum_{i=1}^{N} a_i^2 \, \text{var}(x_i) = \sum_{i=1}^{N} a_i^2 \sigma^2$$

This result is true because we have assumed the rvs to be independent. Comparing this result with Equation 2.91, we observe that all the a_is are the same and, hence,

$$\sum_{i=1}^{N} \left(\frac{1}{N}\right)^2 \sigma^2 = \frac{N}{N^2}\sigma^2 = \frac{1}{N}\sigma^2$$

chapter three

Nonparametric (classical) spectra estimation

Estimating a power spectrum in a long random sequence, $\{x(n)\}$, can be accomplished by, first, creating its autocorrelation function and then taking the Fourier transform (FT). However, there are several problems that appear in establishing power spectra densities. First, the sequence may not be long enough and, sometimes can be very short. Second, the spectral characteristics may change with time. Third, data very often are corrupted with noise. Therefore, the spectrum estimation is a problem that involves estimating $S_x(e^{j\omega})$ from a finite noisy measurement, $x(n)$.

3.1 Periodogram and correlogram spectra estimators

Deterministic signals

If the sequence $\{x(n)\}$ is a **deterministic** real data sequence and has finite energy, e.g.,

$$\sum_{n=-\infty}^{\infty} x^2(n) < \infty \tag{3.1}$$

then the sequence possesses a discrete-time Fourier transform (DTFT):

$$X(e^{j\omega}) = \sum_{n=-\infty}^{\infty} x(n)e^{-j\omega n} \tag{3.2}$$

The inverse DTFT (IDTFT) is (see Problem 3.1.1):

$$x(n) = \frac{1}{2\pi} \int_{-\pi}^{\pi} X(e^{j\omega})e^{j\omega n} d\omega \tag{3.3}$$

where ω is the radians per unit interval.

Note: *The time function is discrete and the transformed one is continuous.*

Note: *The reader should remember that, if the sequence is the result of sampling a continuous signal, the discrete frequency can be easily converted to physical* *frequency rad/s by dividing it by the sampling time T_s. Hence, the frequency range becomes $-\pi/T_s \le \omega \le \pi/T_s$. In addition, we make the following changes in Equation 3.2: $X(e^{j\omega T_s})$, $T_s x(nT_s)e^{-jn\omega T_s}$.*

We can define the **energy spectral density** as follows:

$$S_x(e^{j\omega}) = \left|X(e^{j\omega})\right|^2 \tag{3.4}$$

The Parseval's theorem for real sequence states the following (see Problem 3.2):

$$\sum_{n=-\infty}^{\infty} x^2(n) = \frac{1}{2\pi}\int_{-\pi}^{\pi} S_x(e^{j\omega})\,d\omega \tag{3.5}$$

Note: *This formula states that the total energy in the time domain is equal to the total power in the frequency domain. It does not tell us what part of the time domain energy is distributed in a specific frequency range.*

The autocorrelation of a sequence is defined as follows:

$$r_x(k) = \sum_{n=-\infty}^{\infty} x(n)x(n-k) \tag{3.6}$$

Its DTFT is equal to its power density spectrum (see Problem 3.1.3):

$$\boxed{\sum_{k=-\infty}^{\infty} r_x(k)e^{-j\omega k} = S_x(e^{j\omega})} \tag{3.7}$$

3.1.1   *Random signals*

Periodogram

Stationary random signals, which we will study in this text, have usually finite average power and, therefore, can be characterized by an average power spectral density (PSD). We shall call such a quantity, the PSD. Without loss of generality, the discrete real random sequences, which we will study in this chapter, will assume a zero mean value, e.g.,

$$E\{x(n)\} = 0, \quad \text{for all } n \tag{3.8}$$

The **periodogram** spectral density is defined as follows:

$$\boxed{S_x(e^{j\omega}) = \lim_{N\to\infty} E\left\{\frac{1}{2N+1}\left|\sum_{n=-N}^{N} x(n)e^{-j\omega n}\right|^2\right\}} \tag{3.9}$$

For a finite random sequence, $\{x(n)\}$, the **periodogram spectral estimator** is given by

$$\hat{S}_{px}(e^{j\omega}) = \frac{1}{N}\left|\sum_{n=0}^{N-1} x(n)e^{-j\omega n}\right|^2 = \frac{1}{N}\left|X(e^{j\omega})\right|^2 \qquad (3.10)$$

where $\hat{S}_{px}(e^{j\omega})$ is positive, since it is equal to the square of the absolute value of the DTFT function of $\{x(n)\}$, and it is periodic with period 2π, $-\pi \le \omega \le \pi$. The periodicity is simply shown by introducing $\omega + 2\pi$ in the place of ω and remembering that $\exp(j2\pi) = 1$. The periodogram is an **asymptotic estimator** since as $N \to \infty$ the estimate periodogram converges to the power spectrum of the process. However, the periodogram is **not a consistent estimate** of the power spectrum. This means that the variance of the periodogram does not decrease by increasing the number of terms.

Book MATLAB® function: [s,as,ps]=sspperiodogram(x,L)

```
function[s,as,ps]=sspperiodogram(x,L)
%[s,as,ps]=sspperiodogram(x,L)
%L=desired number of points (bins) of the spectrum;
%x=data in row form; s=complex form of the DFT;
%the directory sspdata.mat contains the file ch3-data1 which
%contains ten thousand elements of two sine signals, a normal
%noise sequence and the signal x of the sum of the three
                %signals;
for m=1:L
    n=1:length(x);
    s(m)=sum(x.*exp(-j*(m-1)*(2*pi/L)*n));
end;
as=((abs(s)).^2/length(x));%as=amplitude spectral
                %density;
ps=(atan(imag(s)./real(s))/length(x));%ps=phase spectrum;

%To plot as or ps we can use the command:
%plot(0:2*pi/L:2*pi-(2*pi/L),as) ; for the phase spectrum
%we change as to ps;
```

To plot **as** or **ps** we can use the command format: `plot(0:2*pi/L:2*pi-(2*pi/L),as);`

The **bias** and **variance** are two basic measures often used to characterize the performance of spectrum estimation. It turns out (see Problem 3.1.4) that the **mean squared error** (MSE) of the estimate is

$$\text{MSE} \triangleq E\{(\hat{S}-S)^2\} = E\{(\hat{S}-E\{\hat{S}\})^2\} + (E\{\hat{S}\}-S)^2 = \text{var}\{\hat{S}\} + \text{bias}\{\hat{S}\}^2 \qquad (3.11)$$

By considering the bias and variance separately, we gain some insight into the source of error and its magnitude. From Equation 3.11, the bias of any estimator $\hat{\theta}$ is given by

$$B(\hat{\theta}) = E\{\hat{\theta}\} - \theta \tag{3.12}$$

where θ is the population value of the parameter. The bias measures the **average** deviation of the estimator from its true value. An estimator is said to be **unbiased** if $B(\hat{\theta}) = 0$. In this case, the MSE becomes equal to variance, and the mean value of the time series is equal to the population mean.

Correlogram

The **correlogram spectral estimator** is based on the formula:

$$\hat{S}_{cx}(e^{j\omega}) = \sum_{m=-(N-1)}^{N-1} \hat{r}_x(m)e^{-j\omega m} \tag{3.13}$$

The **biased** autocorrelation function is given by

$$\hat{r}_x(k) = \frac{1}{N}\sum_{n=k}^{N-1} x(n)x(n-k), \qquad 0 \le k \le N-1 \tag{3.14}$$

The unbiased one is given by the same formula, with the difference, instead of dividing by N the expression is divided by $N - k$. In practice, the biased autocorrelation is used since it produces positive definite matrices and, thus, can be inverted.

To produce the biased correlation function, the following Book MATLAB function may be used:

Book MATLAB function for finding the biased autocorrelation function: [r]=ssp_sample_biased_autoc(x,lg)

```
function[r]=ssp_sample_biased_autoc(x,lg)
%Book MATLAB function:[r]=ssp_sample_biased_autoc(x,lg)
%this function finds the biased autocorrelation function
%with lag from 0 to lg;it is recommended that lg is 20-30% of
%N;
N=length(x);%x=data;lg=lag;
for m=1:lg
    for n=1:N+1−m
        xs(m,n)=x(n−1+m);
    end;
```

```
end;
r1=xs*x';
r=r1'./N;
```

We can also use MATLAB functions to obtain the sample biased and unbiased autocorrelation and cross-correlation. The functions are

```
r_xy=xcorr(x,y);  % x,y are N length vectors; r_xy= a 2N-1
                  %symmetric cross-correlation
                  %vector, in case the vectors do not
                  %have the same length, the shorter
                  %one will be padded with zero; Note:
                  %The correlation sequence is not
                  %divided by N or N-1;
r_xy=xcorr(x,y, 'biased'); %will give the biased cross-
                           %correlation function and the
                           %sequence is divided by N;
r_xy=xcorr(x,y, 'unbised'); %will give the unbiased
                            %autocorrelation function;
```

The Book MATLAB function given below produces the PSD known as the correlogram:

Book MATLAB function for finding the windowed correlogram:
[s,asc,psc]=sspcorrelogram(x,w,lg,L)

```
function[s,asc,psc]=sspcorrelogram(x,w,lg,L)
%[s]=sspcorrelogram(x,w,lg,L);
%x=data with mean zero;w=window(@name,length(2*lg)), see
                  %sspperiodogram
%function and below this function;L=desired number of
                  %spectral points;
%lg=lag number<<N;rc=symmetric autocorrelation function;
r=sspsamplebiasedautoc(x,lg);
rc=[fliplr(r(1,2:lg)) r 0];
rcw=rc.*w';
for m=1:L
  n=-lg+1:lg;
  s(m)=sum(rcw.*exp(-j*(m-1)*(2*pi/L)*n));
end;
asc=(abs(s).^2)/norm(w);%asc=amplitude spectrum;
psc=(atan(imag(s))/real(s))/norm(w);%psc=phase
                  %spectrum;
```

Note: *In case, the reader does not want to use window, he/she must set w=window (@rectwin,length(2*lg)') or w=rectwin(2*lg)'. The same can be done for any other*

*window e.g., w=hamming(2*lg)'. Note the MATLAB transpose sign'. The sign is used because MATLAB gives the window vector in column form.*

General remarks

- In general, the stationary signals have biased correlations that tend to zero as k increases. However, the unbiased autocorrelation may have large values due the division by $N - k$, which takes small values for k close to N.
- The sequence obtained from Equation 3.14 produces a positive semi-definite sequence and, as a consequence, the correlation matrix is also positive semidefinite. This property guarantees that the PSD is positive for all frequencies.
- $\hat{S}_{px}(e^{j\omega}) = \hat{S}_{cx}(e^{j\omega})$ using Equations 3.10 and 3.13 (see Problem 3.1.5).
- The periodogram is **asymptotically unbiased** (see Problem 3.1.6).
- It can be shown that the variance does not go to zero as $N \to \infty$ and, therefore, we conclude that the periodogram **is not a consistent estimator** (see Example 3.1).
- Resolution is proportional to the factor $1/N$. In other words, the resolution improves with the increase in the length of the signal (see Example 3.2).

Example 3.1

Find the periodogram for a white Gaussian (WG) signal with $N = 128$ and 1024 and observe the variance (the variability) of the signal.

Solution

Figure 3.1 shows two periodograms by simulation which indicate that the variance does not decreases with increase in the number of samples. This can also be proved analytically and is found in advanced texts on this subject.   ∎

Example 3.2

Demonstrate the resolution property of the periodogram.

Solution

Figure 3.2 shows two plots of the PSD of a sequence made up of two sinusoids plus a WG noise. The signals were $s = \sin(0.3\pi n) + \sin(0.35\pi n) + 0.5\,\text{randn}(1,64)$ and $s_1 = \sin(0.3\pi n) + \sin(0.35\pi n) + 0.5\,\text{randn}(1,256)$. The two

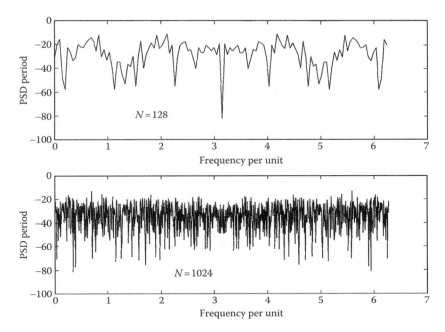

Figure 3.1

frequencies were at 0.9425 and 1.0996 frequency units. The difference between these two frequencies is 0.0571. The resolution for $N = 64$ is $2\pi/64 = 0.0982$ and for $N = 256$ is $2\pi/256 = 0.0245$. From these values and the figure, it is apparent how the resolution property of the periodogram depends on. The effect of the number of samples, N, is also apparent from the figure. ∎

Windowed periodogram

We can define a windowed periodogram as the DTFT of the widowed time function $x(n)w(n)$, where $w(n)$ is any desired window. Hence, the periodogram with temporal windowing is given by

$$\hat{S}_x(e^{j\omega}) = \frac{1}{NP_w}\left|\sum_{n=-\infty}^{\infty} x(n)w(n)e^{j\omega n}\right|^2 \tag{3.15}$$

where

$$P_w = \frac{1}{N}\sum_{n=0}^{N-1}|w(n)|^2 \tag{3.16}$$

where P_w is the average power of the window.

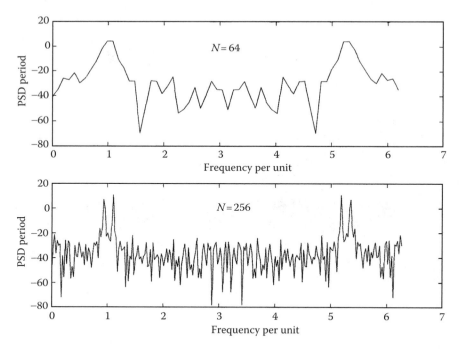

Figure 3.2

The benefit we derive from time windowing is that the spectra are smoother and, sometimes, help differentiate different line spectrums (see Figure 3.3).

Example 3.3

Apply the rectangular and Hamming window to the signal given below and compare their spectrums.

$$s(n) = [0.06\sin(0.2\pi n) + \sin(0.3\pi n)$$
$$+ 0.05(\text{rand}(1,128) - 0.5)]w(n), \quad 0 \le n \le 127$$

Solution

The window provides a trade-off between spectra resolution (main lobe) and spectral masking (side lobe amplitude). In this case, the Hamming window has side lobe at about −45 dB comparing with the rectangular window having a side lobe at about −13 dB. Figure 3.3 shows that, although the side lobe of the rectangular window just about obscures the 0.3π frequency, the Hamming window clearly resolves it. ■

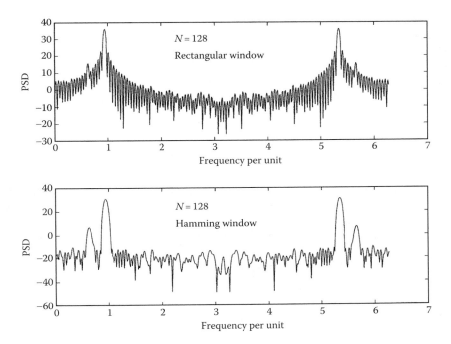

Figure 3.3

It can be shown that the amount of smoothing in the periodogram is deter-mined by the type of window that is applied to the data. Selecting the type of window, we must weight our intention between resolution and smoothing.

The following Book MATLAB function calculates, also the temporal win-dowed data, as the Book MATLAB given in Section 3.1.1, with the difference that with this program the reader can introduce any one of the several win-dows giving in the function without introducing it in the command window. This helps if one wants to repeat the simulation using different windows.

Book MATLAB function for the periodogram with temporal windowed data: [s,as,ps] = sspperiodogramwin(x,win,L)

```
function[s,as,ps]=sspperiodogramwin(x,win,L)
%[s,as,ps]=sspperiodogram(x,win,L)
%window names=hamming,kaiser,hann,rectwin,
%bartlett,tukeywin,blackman,gausswin,
%nattallwin,triang,blackmanharris;
%L=desired number of points (bins) of the spectrum;
%x=data in row form;s=complex form of the DFT;
if (win==2) w=rectwin(length(x));
elseif (win==3) w=hamming(length(x));
elseif (win==4) w=bartlett(length(x));
elseif (win==5) w=tukeywin(length(x));
```

```
elseif (win==6) w=blackman(length(x));
elseif (win==7) w=triang(length(x));
elseif (win==8) w=blackmanharris(length(x));
end;
xw=x.*w';
for m=1:L
    n=1:length(x);
    s(m)=sum(xw.*exp(-j*(m-1)*(2*pi/L)*n));
end;
as=((abs(s)).^2/length(x))/norm(w);%as=amplitude
                        %spectral density;
ps=(atan(imag(s)./real(s))/length(x))/norm(w);%ps=phase
                        %spectrum;
%To plot as or ps we can use the command:
                        %plot(0:2*pi/L:2*pi-(2*pi/L),as);
```

Spectra of some of the windows

Figure 3.4 shows some of the windows and their spectra. Observe that the rectangular window has the narrowest main lobe, but its first side lobe is merely about −13 dB. However, the first side lobe of the Hamming window is about −45 dB.

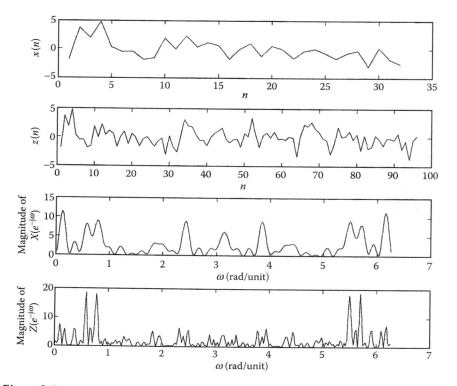

Figure 3.4

Proposed method for better resolution using transformation of the random variables (rvs)

One of the difficulties is that, if the sequence is short (*N* small number), we may not be able to sufficiently resolve frequencies being close together. Since we have at hand one realization and we cannot extract another one from the population in the probability space, we propose to create a pseudosequence from the data in hand by linear transformation of the rvs. Next, we overlap (about 25%) these two series. These series can also be multiplied by a window and then processed.

Note: *The proposed modified method is based on the linear transformation of rvs.*

Example 3.4

Compare the spectrum based on the periodogram and the proposed modified one.

Solution

Figure 3.5 shows the resolutio n capabilities of the sequence which is made up from the original sequence as follows: $y = 0.2*[x \ zeros(1,48)]+0.2*[zeros$

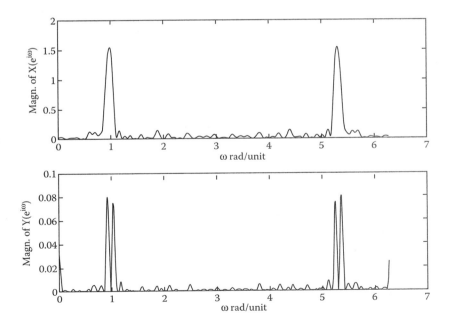

Figure 3.5

(1,48) x];. The original signal, a 64-term series, was
x=sin(0.3*pi*n)+sin(0.32*pi*n)+0.5*
randn(1,64). The following Book MATLAB function
produced the figure.

Book MATLAB function for linearly modified periodogram

```
function[ax,ay,w,y]=ssplinear_modified_periodogram(x)
%Book MATLAB function: [ax,ay,w]:ssplinear_
                              %modified_periodogram(x)
[sx,ax,px]=sspperiodogramwin(x,2,512);
y1=[x zeros(1,48)]+0.2*rand(1,112);
y2=[zeros(1,48) x]+0.2*rand(1,112);
y=(0.2*y1+(y2)*0.2);
[sy,ay,py]=sspperiodogramwin(y,2,512);
w=0:2*pi/512:2*pi-(2*pi/512);
%2 implies rectangular window, see Book MATLAB function
                      %sspperiodogramwin
%for other windows;to plot we must write in the command
                      %window:
%plot(w,20*log10(ax/max(ax)),'k') and similar for the ay;.
```
∎

3.2   *Daniell periodogram*

Daniell suggested that, for a discrete Fourier transform (DFT) periodogram, the sampled PSD $\hat{S}_x(e^{j\omega_i})$ at a particular frequency is found by averaging K points on either side of this frequency. The Daniell formula is

$$\hat{S}_{xD}(e^{j\omega_i}) = \frac{1}{2K+1}\sum_{k=i-K}^{i+K}\hat{S}_x(e^{j\omega_k})$$

(3.17)

This procedure is equivalent to passing the periodogram through a low-pass filter.

3.3   *Bartlett periodogram*

The variance of the periodogram, as given in Equation 3.9, presupposes an ensemble averaging for variance decrease. However, for a single realization the variance does not decreases as $N \to \infty$ (inconsistent estimator) and, hence, there is a need for other approaches to reduce the variance. We can improve the statistical properties of the periodogram by replacing the expectation operator with averaging a set of periodograms. Figure 3.6 shows the ensemble averaging effect on the variance using ten realizations of the PSD only. The

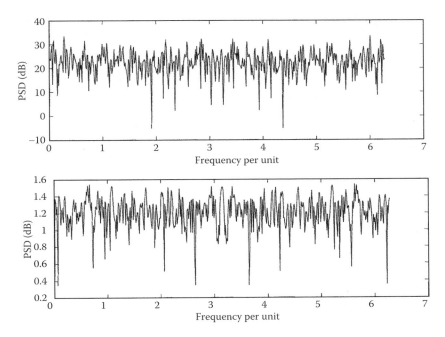

Figure 3.6

variance of one realization is about 55 and the variance of the ensemble is about 2.5. The number of realizations in this case was 30. The reader should compare this figure with Figure 3.1. The following Book MATLAB script file produces Figure 3.6.

Book MATLAB m-File: fig3_3_1

```
%m-file:fig3_3_1
N=256;
for m=1:30
    x(m,1:N)=randn(1,N);
end;
w=0:2*pi/512:2*pi-(2*pi/512);
subplot(2,1,1);plot(w,20*log10(abs(fft...
  (x(10,:),512)))),'k');
subplot(2,1,2);plot(w,20*log10(abs(fft(sum(x,1),...
  512))))/30,'k');
xlabel('Freq. per unit');ylabel('PSD in dB');
```

Bartlett proposed to split the data into *K* nonoverlapping segments, which are assumed to be statistically independent. This condition presupposes that the autocorrelation of each segment decays much faster than the length of the segment. The N samples of the data signal is divided into K nonoverlapping segments of M samples each such that $KM \leq N$. The ith segment will then consists of the samples:

$$x^{(i)}(n) = x(iM + n), \quad n = 0, 1, \ldots, M-1,$$
$$i = \text{segment number} = 0, 1, \ldots, K-1$$

(3.18)

Each of the segments has the spectrum

$$\hat{S}_x^{(i)}(e^{j\omega}) = \frac{1}{M}\left|\sum_{n=0}^{M-1} x(iM + n)e^{-j\omega n}\right|^2$$

(3.19)

and, hence, the Bartlett average periodogram is given by

$$\hat{S}_{xB} = \frac{1}{K}\sum_{i=0}^{K-1}\hat{S}_x^{(i)}(e^{j\omega})$$

(3.20)

See Figure 3.7 for a diagrammatic representation of the above equations defining the Bartlett approach. Since the power spectrum is due to the reduced number of terms, $M \ll N$, comparing to the total available data, the resolution reduces from the order of $1/N$ to the order of $1/M$. This implies

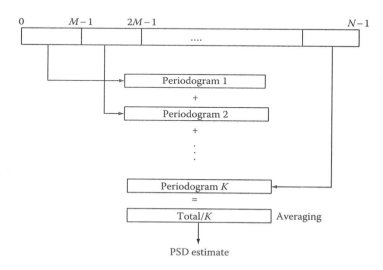

Figure 3.7

that the resolution is reduced by $N/M = K$ comparing to the resolution of the original data set. It can be shown that the variance is reduced by the same amount.

The following Book MATLAB function produces the Bartlett periodogram with a rectangular window:

Book MATLAB function: sspbartlettpsd(x,k,win,L)

```
function[s,as,ps]=sspbartlettpsd(x,k,L)
%x=data;k=number of sections;
%L=number of points desired in the FT domain;
%M=number of points in each section;kM<=N=length(x);
%the number 2 in the text function means rectangular window;
M=floor(length(x)/k);
s=0;
ns=1;
for m=1:k
    s=s+sspperiodogramwin(x(1,ns:ns+M-1),win,L);
    ns=ns+M;
end;
as=((abs(s)/k).^2)/length(x);
ps=(atan(imag(s/k)./real(s/k)))/length(x);
```

Linear transformation modified method

Figure 3.8 is produced, also, to demonstrate the ability of the proposed modified method that uses a linear transformation of the signal at hand to produce a longer pseudorealization of the data and thus improves the resolution. The data for this figure were the following: n=0:127; x=si n(0.3*n*pi)+sin(0.315*n*pi)+0.5*randn(1,128); as indicated above. Figure 3.8a shows the PSD of the 128-long sequence using the Bartlett approach with two sections of 64 elements each. The window for this case was the rectangular window. Figure 3.8b is the same PSD with the difference that a Hamming window is used. The last three rows of the figure present the three proposed modified approaches, with the first column presenting the use of the rectangular window and the second column the use of the Hamming window. The three different Book MATLAB functions for producing the modified spectrums are given below. The figure shows the average PSD of 10 realizations. However, since the amplitude spectrum *as*, for example, is an $R \times 512$ matrix, we can plot only one realization by invoking the command: plot(as(6,:). Hence, in this case the sixth realization, out of R such realizations, is plotted. In the modified cases, the sequences were split into two equal sequences, as was done in the case of Figure 3.8a and b.

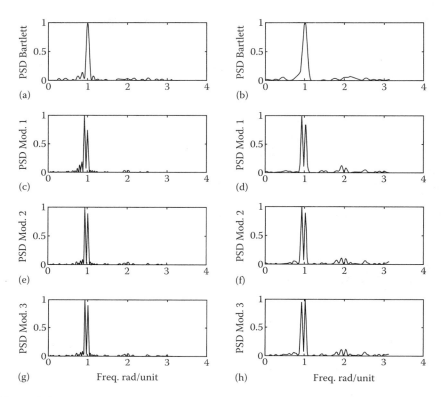

Figure 3.8

Book MATLAB functions for the modified Bartlett spectrum
Modified No. 1

```
function[s,as,ps,apsd]=ssp_aver_mod_ ...
  bartlettpsd1(x,k,win,L,R)
%x=data; k=number of sections; L=number of points
%desired in the FT domain; M=number of points in each
%section;kM<=N=length(x);R=number of realizations;
%win=2 for rectwin, 3 for hamming etc, see
%sspperiodogramwin function;
%s, as and ps are RxL amtrices;apsd is the average spectrum
%of the R realizations;
%if desired we can plot the amplitude spectrum of one
%realization:plot(as(5,1:512));
for r=1:R
```

```
xr(r,:)=([x zeros(1,floor(0.8*length(x)))]+ ...
[zeros(1,floor(0.8*length(x))) x+0.1*rand ...
  (1,length(x))])*.2;
[s(r,:),as(r,:),ps(r,:)]=sspbartlettpsd(xr,k,win,L);
end;
apsd=sum(as,1)/R;
```

Modified No. 2

```
function[s,as,ps,apsd]=ssp_aver_mod_bartlettpsd2 ...
  (x,k,win,L,R)
%x=data; k=number of sections; L=number of points
%desired in the FT domain; M=number of points in each
%section;kM<=N=length(x);R=number of realizations;
for r=1:R
x1(r,:)=[ x*0.5+0.05*randn(1,length(x))...
  x*0.5+0.05*randn(1,length(x))];
[s(r,:),as(r,:),ps(r,:)]=sspbartlettpsd(x1,k,win,L);
end;
apsd=sum(as,1)/R;
```

Modified No. 3

```
function[s,as,ps,apsd]=ssp_aver_mod_bartlettpsd3 ...
  (x,k,win,L,R)
%x=data; k=number of sections; L=number of points
%desired in the FT domain; M=number of points in each
%section;kM<=N=length(x);R=number of realizations;
for m=1:R
x1(m,:)=[x 0.2*x+0.1*randn(1,length(x))];
[s(m,:),as(m,:),ps(m,:)]=sspbartlettpsd(x1,k,win,L);
end;
apsd=sum(as,1);
```

3.4 Blackman–Tukey (BT) method

Because the correlation function at its extreme lag values is not reliable due to the small overlapping of the correlation process, it is recommended to use lag values of about 30%–40% of the total length of the data. The BT estimator is a windowed correlogram and is given by

$$\hat{S}_{BT}(e^{j\omega}) = \sum_{m=-(M-1)}^{M-1} w(m)\hat{r}(m)e^{-j\omega m} \qquad (3.21)$$

where $w(m)$ is the window with zero values for $|m| > M - 1$ and $M \ll N$. This window is known as the **lag window**. The above equation can also be written in the form:

$$\hat{S}_{BT}(e^{j\omega}) = \sum_{m=-\infty}^{\infty} w(m)\hat{r}(m)e^{-j\omega m}$$

$$= \frac{1}{2\pi}\hat{S}_c(e^{j\omega}) * W(e^{j\omega}) = \frac{1}{2\pi}\int_{-\pi}^{\pi}\hat{S}_c(e^{j\tau})W(e^{j(\omega-\tau)})d\tau \qquad (3.22)$$

where we applied the DTFT frequency convolution property (the DTFT of the multiplication of two functions is equal to the convolution of their FTs). Since windows have a dominant and relatively strong main lob, the BT estimator corresponds to a locally weighting average of the periodogram. Although the convolution smoothes the periodogram, it reduces resolution in the same time. It is expected that the smaller the M, the larger the reduction in variance and the lower the resolution. It turns out that the resolution of this spectral estimator is in the order of $1/M$, whereas its variance is in the order of M/N. Observe that the trade-off between resolution (smoothing) and variance depends on the total number of lags (M) we retain.

Book MATLAB function for BT periodogram:
[s,as,ps]=sspBTperiodogram(x,win,per,L)

```
function[s,as,ps]=sspBTperiodogram(x,win,per,L)
%[s,as,ps]=sspbtperiodogram(x,win,L)
%window names=hamming,kaiser,hann,rectwin,
%bartlett,tukeywin,blackman,gausswin,nattallwin,
%triang,blackmanharris;
%L=desired number of points (bins) of the spectrum;
%x=data in row form;s=complex form of the
%DFT;NOTE:per=the percentage
%of points (length(rxt/2))deleted from the correlation
%function
%to decrease the edge effect;
rxt=xcorr(x,'biased');
wn=[zeros(1,floor((length(rxt)*per/2))) ones ...
  (1,length(rxt)-(2*floor((length(rxt)*per/2)))) ...
  zeros(1,floor((length(rxt) *per/2)))];
rx=rxt.*wn;
if (win==2) w=rectwin(length(rx));
elseif (win==3) w=hamming(length(rx));
elseif (win==4) w=bartlett(length(rx));
elseif (win==5) w=tukeywin(length(rx));
```

```
elseif (win==6) w=blackman(length(rx));
elseif (win==7) w=triang(length(rx));
elseif (win==8) w=blackmanharris(length(rx));
end;
rxw=rx.*w';
for m=1:L
    n=1:length(rx);
    s(m)=sum(rxw.*exp(-j*(m-1)*(2*pi/L)*n));
end;
as=((abs(s)).^2/length(x))/norm(w);%as=amplitude
                       %spectral density;
ps=atan((imag(s)./(real(s)+eps))/((length(x))/...
  (norm(w)+eps)));%ps=phase spectrum;
%To plot as or ps we can use the command: plot
%(0:2*pi/L:2*pi-(2*pi/L),as);
```

We propose to increase the window length in steps and then average the spectra. This will produce better resolution but with, somewhat, increase of variance of the spectrum. Figure 3.9 shows clearly the ability to resolve close frequencies by using the proposed averaging method. As it was

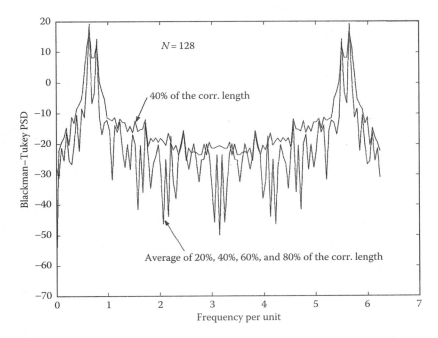

Figure 3.9

anticipated, the variance increases. We observe that using 40% of the correlation, the resolution from the lower peak was about 5 dB, whereas the average procedure gave resolution close to 20 dB.

A second modified method is put forward for the BT method by using linear transformation to rvs. The upper left part of Figure 3.10 shows the original data: $x = \sin(0.3\pi n) + \sin(0.32\pi n) + 0.8$ randn(1,64). The second signal is equal to the original plus a linear transformation of the original: $z = ([x$ zeros(1,48)] + [zeros(1,48) $x])*0.2 + 0.1*$rand(1,112).*hamming(112)';. This signal is plotted in the upper right side of Figure 3.10. The middle section and left part of Figure 3.10 show the BT PSD of the original sequence on a linear scale. The right side of the middle section depicts the BT PSD of the modified sequence on a linear scale. The bottom part of the figure shows the spectrums in the log-linear scale, respectively. For both cases, the percentage of eliminating the trailing correlations terms was 40% of the one-sided correlation or 80% of the symmetric case given by the MATLAB function xcorr(x).

A different modified approach, without overlapping, is given below by the following Book MATLAB program. The sequences and their spectrums are shown in Figure 3.11.

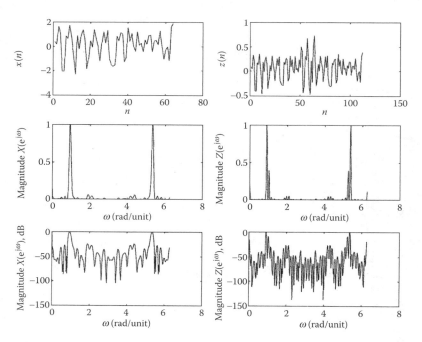

Figure 3.10

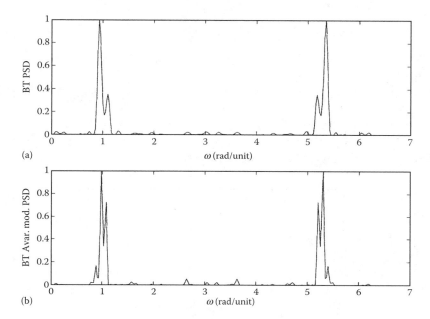

Figure 3.11

Book MATLAB program for modified BT method using random variables transformations

```
n=0:63;
x=sin(0.3*pi*n)+sin(0.32*pi*n)+0.8*randn(1,64);
z=[x*0.5     x*0.25+0.05].*hamming(128)';
[sx,asx,apx]=sspbtperiodogram(x,2,0.4,512);
[sz,,asz,apz]=sspbtperiodogram(z,2,0.4,412);
```

It is interesting to see, by creating an ensemble of modified sequences, if the resolution increases or not. Figure 3.12a shows the BT PSD of the sequence: $x = \sin(0.3*pi*n) + \sin(0.32*pi*n) + 0.8*randn(1,64)$, and Figure 3.12b shows the average (20 sequences) modified BT PSD. Figure 3.12b was created using the following Book MATLAB function:

Book MATLAB function for modified BT method using rvs transformation and averaging: sspav_modifiedbtpsd

```
function[sz,az,pz]=sspav_modified_btpsd(x,win,per,avn,L)
%win=2 implies rectangular, see the function
%sspbtperiodogram
```

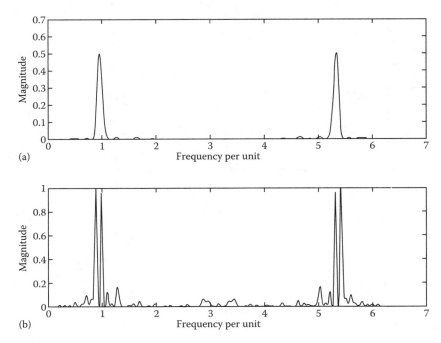

(a)

(b)

Figure 3.12

```
%to identify the appropriate numbers for other windows;per=
%percentage of deleted one sided aucoralation function;
%avn=number of ensemble sequences to be
%averaged;L=number
%of frequence bins;
for m=1:avn
    z1(m,:)=[x*rand*0.5 x*rand*0.2+...
    0.05*(rand-0.5)].*hamming(2*length(x))';
end;
z=sum(z1,1);
[sz,az,pz]=sspbtperiodogram(z,win,per,L);
```

3.5   Welch method

Welch proposed modifications to Bartlett method, as follows: data segments are allowed to overlap and each segment is windowed prior to computing the periodogram. Since, in most practical applications, only a single realization is available, we create smaller sections as follows:

$$x_i(n) = x(iD + n)w(n), \qquad 0 \le n \le M - 1, \quad 0 \le i \le K - 1 \tag{3.23}$$

where
    $w(n)$ is the window of length M
    D is an offset distance
    K is the number of sections that the sequence $\{x(n)\}$ is divided into

Pictorially, the Welch method is shown in Figure 3.13.
    The ith periodogram is given by

$$S_i(e^{j\omega}) = \frac{1}{M}\left|\sum_{n=0}^{M-1} x_i(e^{-j\omega n})\right|^2 \tag{3.24}$$

and the average Welch periodogram is given by

$$\boxed{S(e^{j\omega}) = \frac{1}{K}\sum_{i=0}^{K-1} S_i(e^{j\omega})} \tag{3.25}$$

If $D = M$, then the segments do not overlap and the result is equivalent to Bartlett method with the exception that the segments are windowed.

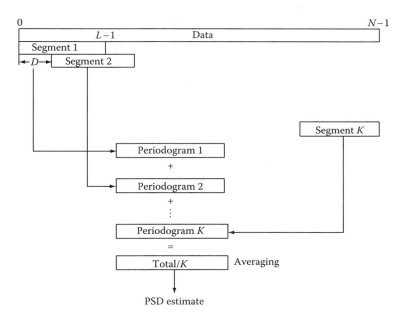

Figure 3.13

Book MATLAB function for Welch periodogram:
[s,as,ps,K]=sspwelchperiodogram(x,w,D,L)

```
function[s,ast,ps,K]=sspwelchperiodogram(x,win,frac, ...
  frac1,L)
%function[as,ps,s,K]=sspwelchperiodogram(x,win,frac,
%frac1,L);
%x=data; M=section length;
%L=number of samples desired in the frequency domain;
%win=2 means rectwin, number 3 means hamming window;
%see sspbtperiodogram function to add more windows if
%you desire;
%frac defines the number of data for each section,
%depending on the data
%length it is recommended the number to vary between 0.2
%and 1;
%frac1 defines the overlapping of the sections, it is
%recommended the frac1 to very from 1 (no overlap) to 0.5
%which means a 50% overlap; M<<N=length(x);
if (win==2) w=rectwin(floor(frac*length(x)));
elseif (win==3) w=hamming(frac*length(x));
end;
N=length(x);
M=floor(frac*length(x));
K=floor(floor((N-M+floor(frac1*M)))/floor(frac1*M));
  %K=number of processings;
s=0;as=0;
for i=1:K
    s=s+fft(x(1,(i-1)*(floor(frac1*M))+1:(i-1)*floor ...
      (frac1*M)+M).*w',L);
    as=as+abs(s);
end;
ast=as/(M*K);%as=amplitude spectral density;
ps=atan(imag(s)./real(s))/(M*K);%phase spectral density;
```

T MATLAB function is given as follows:

```
P=spectrum(x,m,ovelap)%x=data; m=number of points of
                %each section
                %and must be a power of 2;the
                %sections are windowed by a
                %a hanning window;P is a (m/2)x2
                %matrix whose first column is the
                %power spectral density and the
                %second is the 95% confidence
                %interval;
```

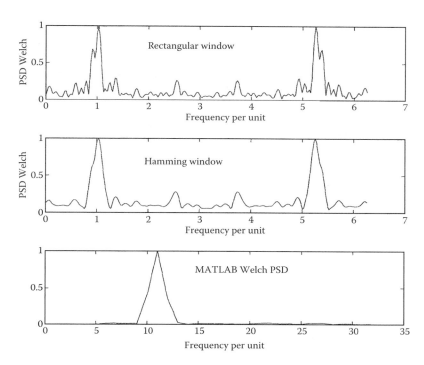

Figure 3.14

```
%if m<256 the spectrum will be given
%for 256 points; overlap is
%the number of points which will
%overlap, must be less than m;
```

Figure 3.14 was created as follows: $n = 0{:}63$; $x = \sin(0.5\ast pi\ast n) + \sin(0.32\ast pi\ast n) + 0.5\ast randn(1,64)$; Figure 3.14a is produced using the Book MATLAB function: **sspwelchperiodogram(x,win,frac,frac1,L)** with a rectangular window. Figure 3.14b is produced using the same Book MATLAB function with the Hamming window. Figure 3.14c is produced using the MATLAB function: **spectrum(x,m,overlap)**.

3.6 Proposed modified methods for Welch periodogram

Modified method using different types of overlappings

It is evident from Figure 3.13 that, if the lengths of the sections are not long enough, frequencies close together cannot be differentiated. Therefore, we propose a procedure, defined as **symmetric modified Welch method**, and its

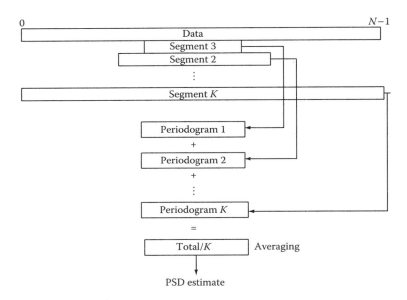

Figure 3.15

implementation is shown in Figure 3.15. Windowing of the segments can also be incorporated. This approach and the rest of the proposed schemes have the advantage of progressively incorporating longer and longer segments of the data and thus introducing better and better resolution. In addition, due to the averaging process, the variance decreases and smoother periodograms are obtained but not as smooth as the Welch method. It is up to the reader to decide between smoothness of the periodogram and resolution of frequencies. Figure 3.16 shows another proposed method, which is defined as the asymmetric modified Welch method. Figure 3.17 shows another suggested approach for better resolution and reduced variance. The procedure is based on the method of prediction and averaging. This proposed method is defined as the **symmetric prediction modified Welch method**. This procedure can be used in all the other forms, e.g., nonsymmetric. The above methods can also be used for spectral estimation if we substitute the word periodogram with the word correlogram.

Figure 3.18a shows data created by the equation:

$$x(n) = \sin(0.3\pi n) + \sin(0.32\pi n) + \text{rand}(1, 128)$$

$$(3.26)$$

and for 128 sample points. Figure 3.18b shows the Welch periodogram result using the MATLAB function: **p=spectrum(x,64,32);**. This function creates a

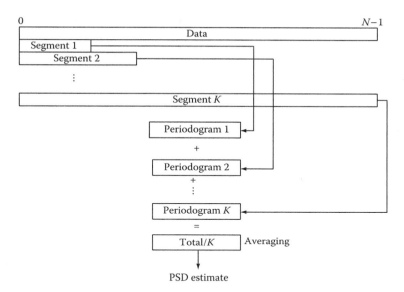

Figure 3.16

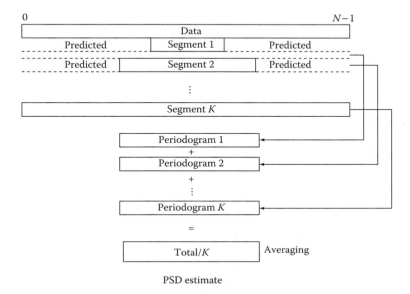

Figure 3.17

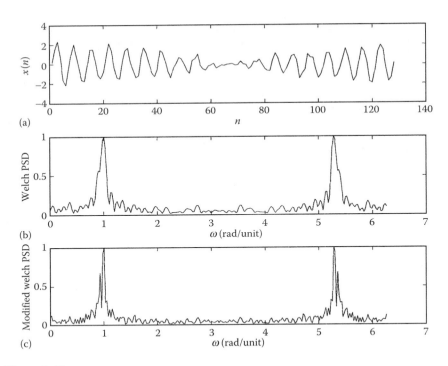

Figure 3.18

spectrum with 256 frequency bins and uses Hanning window for each of the segments. Figure 3.18c shows the proposed asymmetric modified Welch method. Our proposed method gives much better resolution with a very small increase in variance.

Modified Welch method using rvs transformation

Figure 3.19a shows the result of the Welch method for the signal $x = \sin(0.3\pi n) + \sin(0.315\pi n) + \text{randn}(1,128)$, for $n = 0, 1, \ldots, 127$. The signal was split into two sections with 50% overlap. In Figure 3.19b, the following transformed signal was used by incorporating a linear transformation of the rvs: $z = [x\ 0.05^*x + 0.05^*\text{randn}(1,128)]$. For this signal, the Welch method was used for splitting the series into two parts and using 50% overlapping. The figure was produced using the following Book MATLAB program:

Book MATLAB program for comparing the Welch method with the proposed modification by using transformation of the random variables

```
>>n=0:127;
>>x=sin(0.3*pi*n)+sin(0.315*pi*n)+0.5*randn(1,128);
>> [sx,as ps,K]=sspwelchperiodogram(x,2,0.5,0.5,512);
   % 50% overlap;
```

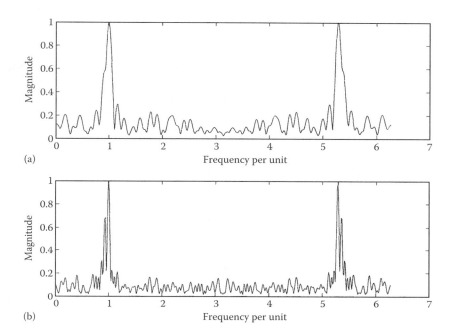

(a)

(b)

Figure 3.19

```
>> plot(w,as/max(as),'k')
>> z=[x 0.1*x+0.05*randn(1,128)];
>> [sz,asz psz,Kz]=sspwelchperiodogram(z,2,0.5,0.5,512);
>> plot(w,asz/max(asz),'k')
```

Problems

3.1.1 Verify Equation 3.3.

3.1.2 Verify Equation 3.5.

3.1.3 Verify Equation 3.7.

3.1.4 Verify Equation 3.11.

3.1.5 Verify the relation $\hat{S}_{px}(e^{j\omega}) = \hat{S}_{cx}(e^{j\omega})$ using standard biased autocorrelation sequences.

3.1.6 Show that the periodogram is asymptotically unbiased.

3.1.7 Create a signal with two sinusoids close in frequency and length 512. Find the periodogram and the correlogram for this signal. Next, reduce its length by 128, 256, and 284 elements and do the same as above. In each case observe the results. Next, repeat the whole procedure; but before finding their spectrums, multiply the data with the Hamming window and compare results.

3.2.1 Create a periodogram and apply Daniell's method for $L = 1, 2, 3$ and observe the resulting spectrums.

3.3.1 Find the bias and the variance of the Bartlett periodogram.

3.3.2 Do the following and observe the results: (1) create a signal of two close in frequency sinusoids plus white noise with $N = 512$ values. (2) Overlay the plots of 50 periodograms with rectangular window and then average. (3) Overlay the plots of 50 Bartlett estimates with $I = 4$ (four sections, $M = 128$) and then average. (4) Overlay the plots of 50 Bartlett estimates with $I = 8$ (eight sections, $M = 64$) and then average.

3.3.3 Repeat the modified Bartlett method as it was proposed and vary the constants in the transformation of the rvs, lengths of data, window types, and observe the results.

3.4.1 Use the BT periodogram function and observe the spectrums for different windows, different variances of the white additive noise, and different sinusoids having close frequencies or not.

3.5.1 Use the Welch approach with different variances of the noise, different distance between frequencies of sinusoids, different windows, and different lengths of segments.

Solutions, suggestions, and hints

3.1.1 Multiply Equation 3.2 by $e^{j\omega m}$ and integrate in the range $-\pi \le \omega \le \pi$. Hence, $\int_{-\pi}^{\pi} X(e^{j\omega})e^{j\omega m}d\omega = \sum_{n=-\infty}^{\infty} x(n)\int_{-\pi}^{\pi} e^{-j\omega(n-m)}d\omega = 2\pi x(n)$ since the integral is zero for $n \ne m$ (an integer) and equal to 2π for $n = m$.

3.1.2 $(1/2\pi)\int_{-\pi}^{\pi} S_x(e^{j\omega})d\omega = (1/2\pi)\int_{-\pi}^{\pi}\sum_{n=-\infty}^{\infty}\sum_{m=-\infty}^{\infty}x(n)x(m)e^{-j\omega(n-m)}d\omega$

$= \sum_{n=-\infty}^{\infty}\sum_{m=-\infty}^{\infty}x(n)x(m)\left[(1/2\pi)\int_{-\pi}^{\pi}e^{-j\omega(n-m)}d\omega\right] = \sum_{n=-\infty}^{\infty}x^2(n);$

$(1/2\pi)\int_{-\pi}^{\pi}e^{-j\omega(n-m)}d\omega = \delta(n-m)$ (Kronecker delta) which is equal to 1 if $m = n$ and 0 if $m \ne n$. Hence, in the expansion of the summations only the terms with square elements of $x(n)$s remain.

3.1.3 $\sum_{k=-\infty}^{\infty}r_x(k)e^{-j\omega k} = \sum_{k=-\infty}^{\infty}\sum_{n=-\infty}^{\infty}x(n)x(n-k)e^{-j\omega n}e^{j\omega(n-k)}$

$= \left[\sum_{n=-\infty}^{\infty}x(n)e^{-j\omega n}\right]\left[\sum_{s=-\infty}^{\infty}x(s)e^{-j(-\omega)s}\right] = X(e^{j\omega})X(e^{-j\omega})$

$= \left|X(e^{j\omega})\right|^2 = S_x(e^{j\omega})$

3.1.4 MSE $\triangleq E\{(\hat{S}-S)^2\} = E\{(\hat{S}-E\{\hat{S}\}+E\{\hat{S}\}-S)^2\}$

$= E\{(\hat{S}-E\{\hat{S}\})^2\} + (E\{\hat{S}\}-S)^2 + 2E\{(\hat{S}-E\{\hat{S}\})(E\{\hat{S}\}-S)\}$

$= E\{(\hat{S}-E\{\hat{S}\})^2\} + (E\{\hat{S}\}-S)^2 + 2[E\{\hat{S}E\{\hat{S}\} - E\{\hat{S}\}E\{\hat{S}\} - \hat{S}S + E\{\hat{S}\}S\}]$

$= \text{var}\{\hat{S}\} + \text{bias}\{\hat{S}\} + 2[E\{\hat{S}\}E\{\hat{S}\} - E\{\hat{S}\}E\{\hat{S}\} - E\{\hat{S}\}E\{\hat{S}\} + E\{\hat{S}\}E\{\hat{S}\}]$

$= \text{var}\{\hat{S}\} + \text{bias}\{\hat{S}\}$

3.1.5 $y(n) = \left(1/\sqrt{N}\right)\sum_{m=0}^{N-1} h(m)v(n-m)$, $h() = $ constant, $v() = $ white noise with variance 1 (1)

$$Y(e^{j\omega}) = \left(1/\sqrt{N}\right)\sum_{m=0}^{N-1} y(m)e^{-j\omega m} \qquad (2)$$

$$E\{v(n)v(m)\} = r_v(k) = 1\delta(k), \ k = m - n;$$

$$V(e^{j\omega}) = \left(1/\sqrt{N}\right)\sum_{k=0}^{N-1} r_v(k)e^{-j\omega k} = \sum_{k=0}^{N-1}\delta(k)e^{-j\omega k} = 1 \qquad (3)$$

from 2.46, $S_y(e^{j\omega}) = \left|H(e^{j\omega})\right|^2 = \hat{S}_{py}(e^{j\omega})$ (4)

$$r_y(k) = E\{y(n)y(n-k)\} = \frac{1}{N}\sum_{p=0}^{N-1}\sum_{s=0}^{N-1} h(p)h(s)E\{v(n-p)v(n-k-s)\}$$

$$= \frac{1}{N}\sum_{p=0}^{N-1}\sum_{s=0}^{N-1} h(p)h(s)\delta(k+s-p) = \frac{1}{N}\sum_{p=0}^{N-1} h(p)h(p-k) = r_h(k),$$

for $0 \le k \le N-1$ (5)

Using the DTFT of (5), we find $S_y(e^{j\omega}) = S_h(e^{j\omega}) \triangleq \left|H(e^{j\omega})\right|^2$ and, hence, from (4) we obtain the desired results.

3.1.6 To show that the periodogram is asymptotically unbiased we must show that the mean value of the correlation tend to zero as $N \to \infty$. Hence, from Equation 3.14 we have

$$E\{\hat{r}_x(k)\} = \frac{1}{N}\sum_{n=k}^{N-1} E\{x(n)x(n-k)\} = \frac{1}{N}\sum_{n=k}^{N-1}\hat{r}_x(k) = \hat{r}_x(k)\frac{1}{N}(N-k)$$

$$= (1 - \frac{k}{N})\hat{r}_x(k) \qquad (1)$$

Taking into consideration the correlation symmetry, we write

$$E\{\hat{r}_x(k)\} = \begin{cases} \left(1 - \left(|k|/N\right)\right)\hat{r}_x(k), & |k| \le N \\ 0, & \text{elsewhere} \end{cases}, \qquad w_B = \begin{cases} \left(1 - \left(|k|/N\right)\right), & |k| \le N \\ 0, & \text{elsewhere} \end{cases}$$

= Bartlett window

Taking the ensemble average of the periodogram we obtain:

$$E\{\hat{S}_{px}(e^{j\omega})\} = E\left\{\sum_{k=-N+1}^{N-1}\hat{r}_x(k)e^{-j\omega k}\right\} = \sum_{k=-N+1}^{N-1} E\{\hat{r}_x(k)\}e^{-j\omega k} = \sum_{k=-\infty}^{\infty} r_x(k)w_B e^{-j\omega k}$$

Since the ensemble average of the periodogram (DTFT) is equal to DTFT of the product $r_x(k)w_B(k)$, the product property of DTFT gives $E\{\hat{S}_{px}(e^{j\omega})\} = (1/2\pi)S_x(e^{j\omega}) * W_B(e^{j\omega})$ where $W_B(e^{j\omega}) = $ [sin $(N\omega/2)/$sin $(\omega/2)]^2/N$ is the DTFT of the Bartlett window. Since the DTFT of the Bartlett window approaches a delta function as N approaches infinity, the convolution becomes $\lim_{N\to\infty} E\{\hat{S}_{px}(e^{j\omega})\} = S_x(e^{j\omega})$.

3.3.1 $E\{\hat{S}_x(e^{j\omega})\} = E\left\{\left[\sum_{n=-\infty}^{\infty} x(n)w_R(n)e^{-j\omega n}\right]\left[\sum_{m=-\infty}^{\infty} x(m)w_R(m)e^{-j\omega m}\right]\right\}$

$$= E\left\{\sum_{m=-\infty}^{\infty}\sum_{n=-\infty}^{\infty} x(n)x(m)w_R(m)w_R(n)e^{-j(n-m)\omega}\right\}$$

$$= \sum_{m=-\infty}^{\infty}\sum_{n=-\infty}^{\infty} r_x(n-m)w_R(m)w_R(n)e^{-j(n-m)\omega}$$

$$= \sum_{k=-\infty}^{\infty}\sum_{n=-\infty}^{\infty} r_x(k)w_R(m)w_R(n-k)e^{-jk\omega} \quad \text{(where we changed the}$$

variables by substituting $k = m - n$)

$$= \sum_{k=-\infty}^{\infty} r_x(k)\left[\sum_{n=-\infty}^{\infty} w_R(n)w_R(n-k)\right]e^{-jk\omega} = \sum_{k=-\infty}^{\infty} r_x(k)w_B(k)e^{-j\omega k} \quad (1)$$

$w_B(k) = \text{Bartlett window} = w_R(k) \otimes w_R(k)(\text{correlation}) = w_R(k) * w_R(-k)$ (2)

But the FT of (2) is equal to the multiplication of the FTs of the rectangular windows or

$$W_B(e^{j\omega}) = \left|W_R(e^{j\omega})\right|^2$$
$$(3)$$

$W_R(e^{j\omega}) = [\sin(N\omega/2)/\sin(\omega/2)]e^{-j(N-1)\omega/2}$ (4) (this can be found using the finite number geometric series) But (1) indicates the FT of the multiplication of two functions and, hence, the FT of the Bartlett periodogram of each section is

$$E\{\hat{S}_x(e^{j\omega})\} = E\{\hat{S}_x^{(i)}(e^{j\omega})\} = (1/2\pi)S_x(e^{j\omega}) * W_B(e^{j\omega}) \quad (5)$$

But the FT of the Bartlett window tends to a delta function for large number of terms and hence it is asymptotically unbiased. It can be shown that the variance of the periodogram is given by (for large number of terms) $\text{var}\{\hat{S}_x(e^{j\omega})\} = S_x^2(e^{j\omega})$ (an inconsistent estimator since the variance does not go to zero). Therefore,

$$E\{\hat{S}_{xB}(e^{j\omega})\} = (1/I)E\{\hat{S}_x^{(i)}(e^{j\omega})\} \approx (1/I)S_x^2(e^{j\omega})$$

which tends to zero as the number of segments increase (the Bartlett periodogram is an asymptotically consistent estimator).

chapter four

Parametric and other methods for spectra estimation

In the previous chapter, we presented the classical (nonparametric) spectra estimation. The procedure was to find the autocovariance function (ACF) of the data and then take the discrete-time Fourier transform (DTFT) of the ACF. In this chapter, we will consider the **parametric** approach that is based on model parameters rather than on ACF. These models include the autoregressive (AR) model, the moving average (MA) model, and the autoregressive-moving average (ARMA) model. It is assumed that the data are the result of the output of one of these systems with input white noise having finite variance. The parameters and the noise variance will be determined as described in this chapter.

The parametric approach has been devised to produce better spectral resolution and better spectra estimation. However, we must be careful in using this method since the degree of improvement in resolution and spectral fidelity is solely determined by the appropriateness of the selected model.

Primarily, we will consider **discrete spectra** (sinusoidal signals) embedded in white noise. The motivation for studying parametric models for spectrums estimation is based on the ability to achieve better power spectral density (PSD) estimation, assuming that we incorporate the appropriate model. Furthermore, in the nonparametric case the PSD was obtained from windowed set of data or autocorrelation sequence (ACS) estimates. The unavailable data, in both the sequences and ACS, imposed the unrealistic assumption that the data outside the windows have zero values. If some knowledge about the underlined process, which produces the sequence, is present then the extrapolation to unavailable data outside the window is more realistic process than in the nonparametric case. This brings up the idea that the window is not more needed and, as a consequence, less distortion may occur to the spectrum.

4.1   AR, MA, and ARMA models

The difference equation that describes the general model is (see also Chapter 2):

$$y(n) = -\sum_{m=1}^{p} a(m)y(n-m) + \sum_{m=0}^{q} b(m)v(n-m) \tag{4.1}$$

For simplicity and without loss of generality, we will assume that $b(0)$ is equal to 1. In the above equation, $y(n)$ represents the output of the system, and $v(n)$, the input white noise to the system. This is an innate-type of input to the system. Any external noise that is present must be separately added to the system. If we consider the impulse response of the system, the above equation takes the form:

$$y(n) = \sum_{m=0}^{\infty} h(n)v(n-m) \tag{4.2}$$

For causal systems, we have the relation $h(m) = 0$ for $m < 0$. If we take the z-transform of both sides of the above equations, we obtain (remember that the z-transform of the convolution of two sequences is equal to the multiplication of their z-transforms):

$$\frac{Y(z)}{V(z)} = \frac{B(z)}{A(z)}, \quad \frac{Y(z)}{V(z)} \triangleq H(z) \quad \text{or} \quad H(z) = \frac{B(z)}{A(z)} \tag{4.3}$$

We will assume that all the roots and zeros of the transfer function $H(z)$ are within the unit circle so that the systems are **stable**, **causal**, and **minimum phase**. The expanded forms of the above polynomials are

$$A(z) = 1 + \sum_{m=1}^{p} a(m)z^{-m}, \quad B(z) = 1 + \sum_{m=1}^{q} b(m)z^{-m}, \quad H(z) = 1 + \sum_{m=1}^{\infty} h(m)z^{-m} \tag{4.4}$$

The power spectrum of the output is given by (see Equation 2.44):

$$S_y(z) = S_v(z)H(z)H(z^{-1}) = S_v(z)\frac{B(z)B(z^{-1})}{A(z)A(z^{-1})} \tag{4.5}$$

This equation represents the ARMA model for the time series $y(n)$ when the input is a zero mean white noise. The ARMA PSD is obtained by setting $z = e^{j\omega}$ in Equation 4.5 yielding

$$\boxed{S_y(e^{j\omega}) = \sigma_v^2 \frac{B(e^{j\omega})B(e^{-j\omega})}{A(e^{j\omega})A(e^{-j\omega})} = \sigma_v^2 \frac{\left|B(e^{j\omega})\right|^2}{\left|A(e^{j\omega})\right|^2}} \tag{4.6}$$

where σ_v^2 is the variance of the white noise that entirely characterizes the PSD of the process $y(n)$. The polynomials in the above equation are

$$A(e^{j\omega}) = 1 + \sum_{m=1}^{p} a(m)e^{-j\omega m}, \quad B(e^{j\omega}) = 1 + \sum_{m=1}^{q} b(m)e^{-j\omega m} \tag{4.7}$$

When the $a(m)$s are zero, the system is an MA one, and when the $b(m)$s are zero, the system is an AR. In this chapter, we shall concentrate on the

AR-type systems since these systems provide the sharpest peaks in the spectrum at the corresponding sinusoidal frequencies and will provide us with the fundamentals of dealing with the processing of random discrete signals. The AR system is characterized in the time and frequency domain by the following two equations, respectively:

$$y(n) = -\sum_{m=1}^{p} a(m)y(n-m) + v(n) \tag{4.8}$$

$$S_y(e^{j\omega}) = \frac{\sigma_v^2}{\left|1 + \sum_{m=1}^{p} a(m)e^{j\omega m}\right|^2} \tag{4.9}$$

For a causal system, the ARMA and AR correlation equations are (see Problem 4.1.1):

$$r_y(k) + \sum_{m=1}^{p} a(m)r_y(k-m) = \sigma_v^2 \sum_{m=0}^{q} b(m)h(m-k) \tag{4.10}$$

Since, in general, $h(k)$ is a nonlinear function of the $a(m)$s and $b(m)$s coefficients and we impose the causality condition, $h(s) = 0$ for $s < 0$, Equation 4.10 for $k \geq q + 1$ becomes

$$r_y(k) + \sum_{m=1}^{p} a(m)r_y(k-m) = 0, \quad k > q \tag{4.11}$$

which is the basis for many estimators of the AR coefficients of the AR process. For the values $k \geq q + 1$ and the restriction of $m \leq q$, the impulse response of a causal system is zero.

4.2   Yule–Walker (YW) equations

Equations 4.10 and 4.11 indicate a linear relationship between the correlation coefficients and the AR parameters. From Equation 4.10, for $k = 0$, we obtain the relation:

$$r_y(0) + \sum_{m=1}^{p} a(m)r_y(0-m) = \sigma_v^2 \sum_{m=0}^{0} b(m)h(m-0) = \sigma_v^2 \tag{4.12}$$

The last equality of the above equation is true since $b(0) = h(0) = 1$ (see Equation 4.4) and any other $b(m)$ is zero for an AR model to be true. The expansion of Equation 4.11 (for an AR model $q = 0$) in connection with Equation 4.12, we obtain the following set of equations:

$$
\begin{bmatrix}
r_y(0) & r_y(-1) & \cdots & r_y(-p) \\
r_y(1) & r_y(0) & \cdots & r_y(-p+1) \\
\vdots & \vdots & & \vdots \\
r_y(p) & r_y(p-1) & \cdots & r_y(0)
\end{bmatrix}
\begin{bmatrix}
1 \\
a(1) \\
\vdots \\
a(p)
\end{bmatrix}
=
\begin{bmatrix}
\sigma_v^2 \\
0 \\
\vdots \\
0
\end{bmatrix}
\tag{4.13}
$$

The above equations are known as the **Yule–Walker** or the **normal** ones. For additional information about vectors and matrices operations see Appendix B. These equations form the basis of many AR estimate methods. The above matrix is Toeplitz because $r_y(k) = r_y(-k)$. If the correlation coefficients are known, the lower part of the above equation can be written as

$$
\begin{bmatrix}
r_y(0) & \cdots & r_y(-p+1) \\
\vdots & & \vdots \\
r_y(p-1) & \cdots & r_y(0)
\end{bmatrix}
\begin{bmatrix}
a(1) \\
a(2) \\
\vdots \\
a(p)
\end{bmatrix}
= -
\begin{bmatrix}
r_y(1) \\
\vdots \\
r_y(p)
\end{bmatrix}
\tag{4.14}
$$

In matrix form, the above equation can be written as

$$
\mathbf{R}_{y(p)}\mathbf{a} = -\mathbf{r}_{y(p)} \text{ and the solution is } \mathbf{a} = -\mathbf{R}_{y(p)}^{-1}\mathbf{r}_{y(p)}
\tag{4.15}
$$

To find σ_v^2, we introduce the $a(n)$s, just found, in the first row of Equation 4.13. It can be shown that the correlation matrix in Equation 4.13 is positive definite and, hence, it has a unique solution.

Equation 4.13 can also be written in the compact matrix form as follows:

$$
\mathbf{R}_{y(p+1)}
\begin{bmatrix}
1 \\
\mathbf{a}
\end{bmatrix}
=
\begin{bmatrix}
\sigma_v^2 \\
0
\end{bmatrix}
\tag{4.16}
$$

Note: *We must have in mind that the true correlation elements {r(k)} are replaced with the sample correlation elements {r̂(k)}.*

Example 4.1

Find the spectrum using the YW equations for the sequence produced by the infinite impulse response (IIR) filter:

$$
y(n) = 1.3847\,y(n-1) - 1.5602\,y(n-2) + 0.8883\,y(n-3) - 0.4266\,y(n-4) + v(n)
$$

where $v(n)$ is a white Gaussian (WG) noise with zero mean and unit variance values. Use 30th and 60th AR estimating models. Produce a sequence with 256 elements and add the following two different sinusoids:

$$s = 2\sin(0.3\pi n) + 2\sin(0.32\pi n), \quad s = 4\sin(0.3\pi n) + 4\sin(0.32\pi n)$$

Solution

To produce Figure 4.1, we used the AR model given above where $v(n)$ is a WG noise with variance 1 and mean value 0. To the above data and for Figure 4.1a and b, the following signal was added: $s(n) = 2\sin(0.3n\pi) + 2\sin(0.32n\pi)$. From Figure 4.1a and b, we observe that as the order of AR estimating model increases, 30 and 60, respectively, the resolution increases. We can definitely identify the two frequencies, although some additional splitting (possible additional line

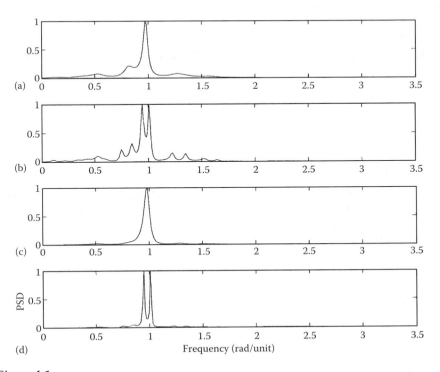

Figure 4.1

spectra) takes place. Figure 4.1c and d was produced with the same data and the same orders, respectively, but with the following signal: $s(n) = 4\sin(0.3n\pi) + 4\sin(0.32n\pi)$. For this case, the separation of the two frequencies is clearly defined. Comparing Figure 4.1a and b with Figure 4.1c and d, we observe that as the signal-to-noise ratio increases, identification of the line spectrum improves considerably. The total number of data in the sequence was 256 and the correlation lag was 128 units. ∎

The following function will produce the estimated AR spectrum:

Book MATLAB® function: sspARpsd1(x,lg,ord,L)

```
function[arft,w,r]=sspARpsd1(x,lg,ord,L)
%x=output of an IIR with input white Gaussian
%noise of variance 1 and mean zero;a(ord)=coefficients
%with order equal to ord;lg=number of lags for the
%autocorralation
%function and MUST be at least one larger than the ord;
%ord=order of the AR model; L=desired bins in the
%spectrum;
%on the command window the plot(w,arft) will plot the
%PSD;
r=sasamplebiasedautoc(x,lg);
m=1:ord;
R=toeplitz(r(1,1:ord));
a=-inv(R)*r(1,2:ord+1)';
w=0:2*pi/L:2*pi-(2*pi/L);
arft=(1./abs((1+(exp(-j*w'*m)*a)).^2))/lg;
```

Figure 4.2 has been produced using the Book MATLAB function given below. Figure 4.2a gives three realizations of the spectrum, and Figure 4.2b is the result of averaging 20 realizations. The order of the AR model estimator was 60, the number of data was 256, and the lag number for the autocorrelation function was 128.

Book MATLAB function: sspARpsd_realiz1 (lg,ord,K,L,N)

```
function[w,r,arft]=sspARpsd_realiz1(lg,ord,K,L,N)
%x=output of an IIR with input white Gaussian
%noise of variance 1 and mean zero;a(ord)=coefficients
%with order equal to ord;lg=number of lags for the
%autocorralation
```

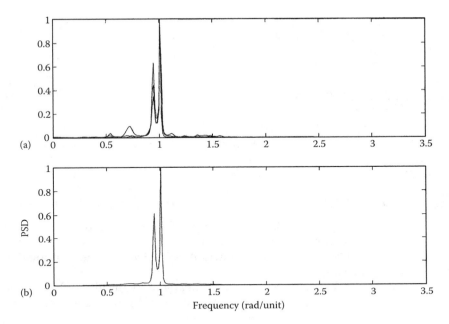

Figure 4.2

```
%function and MUST be at least one larger than the ord;
%ord=order of the AR model; L=desired bins in the
%spectrum power of 2^n;
%on the command window the plot(w,arft) will plot the
%PSD;
%K=number of realizations; N=length of IIR output
%sequence (vector x);
%arft=a LxK matrix with each column representing of the
%PSD for each
%realization, to plot the average you write:
%plot(w,sum(arft,2)), the
%number 2 indicates the summation of columns; the reader
%can easily change
%the signal and the IIR filter to fit the desired
%outputs;
  for k=0:K-1
      for n=0:N-1
      x(4)=0;x(3)=0;x(2)=0;x(1)=0;
      x(n+5)=1.3847*x(n+4)-1.5602*x(n+3)+0.8883*x(n+2)-...
      0.4266*x(n+1)+randn;
end;
```

```
q=0:N-1;
s=4*sin(0.3*pi*q)+4*sin(0.32*pi*q);
x1=x(1,5:N+4)+s;
r=sasamplebiasedautoc(x1,lg);
m=1:ord;
R=toeplitz(r(1,1:ord));
a=-inv(R)*r(1,2:ord+1)';
w=0:2*pi/L:2*pi-(2*pi/L);
arft(:,k+1)=(1./abs((1+(exp(-j*w'*m)*a)).^2))/lg;
end;
```

4.3   Least-squares (LS) method and linear prediction

An alternative approach would be to perform a LS minimization with respect to the linear prediction coefficients. There exist two types of LS estimates, the forward and backward linear prediction estimates, and the combination of forward and backward linear prediction.

Assume the N-data sequence $y(1)$, $y(2)$, ..., $y(N)$ to be used to estimate the pth order AR parameters. The forward **linear prediction** data $\hat{y}(n)$ based on the previously given data $\{y(n-1)\ y(n-2)\ \cdots\ y(n-p)\}$ is

$$\hat{y}(n) = -\sum_{i=1}^{p} a(i)y(n-i) = -\mathbf{y}^{\mathrm{T}}(n)\mathbf{a},$$
$$\mathbf{y}(n) = \begin{bmatrix} y(n-1) & y(n-2) & \cdots & y(n-p) \end{bmatrix}^{\mathrm{T}}, \tag{4.17}$$
$$\mathbf{a} = \begin{bmatrix} a(1) & a(2) & \cdots & a(p) \end{bmatrix}$$

Therefore, the **prediction error** is

$$e_{p}(n) = y(n) - \hat{y}(n) = y(n) + \sum_{i=1}^{p} a(i)y(n-i) = y(n) + \mathbf{y}^{\mathrm{T}}(n)\mathbf{a} \tag{4.18}$$

The vector $\mathbf{a}$, that minimizes the prediction error variance $\sigma_p^2 \triangleq E\{|e(n)|^2\}$, is the desired AR coefficients. From Equation 4.18, we obtain (see Problem 4.3.1):

$$\sigma_p^2 \triangleq E\{|e(n)|^2\} = E\left\{\left[y(n) + \mathbf{a}^{\mathrm{T}}\mathbf{y}(n)\right]\left[y(n) + \mathbf{y}^{\mathrm{T}}(n)\mathbf{a}\right]\right\}$$
$$= r_y(0) + \mathbf{r}_{y(p)}^{\mathrm{T}}\mathbf{a} + \mathbf{a}^{\mathrm{T}}\mathbf{r}_{y(p)} + \mathbf{a}^{\mathrm{T}}\mathbf{R}_{y(p)}\mathbf{a} \tag{4.19}$$

where $\mathbf{r}_{y(p)}$ and $\mathbf{R}_{y(p)}$ are defined in Equation 4.15. The vector $\mathbf{a}$ that minimizes Equation 4.19 (see Problem 4.3.2) is given by

$$\boxed{\mathbf{a} = -\mathbf{R}_{y(p)}^{-1}\mathbf{r}_{y(p)}} \tag{4.20}$$

which is identical to Equation 4.15 and was derived from the YW equations. The minimum prediction error is found to be (see Problem 4.3.3):

$$\sigma_p^2 \triangleq E\{|e(n)|^2\} = y(0) - \mathbf{r}_{y(p)}^T \mathbf{R}_{y(p)}^{-1} \mathbf{r}_{y(p)} \tag{4.21}$$

The LS AR estimation method approximates the AR coefficients found by Equation 4.20, using a finite sample instead of an ensemble, by minimizing the finite-sample cost function:

$$J(\mathbf{a}) = \sum_{n=N_1}^{N_2} |e(n)|^2 = \sum_{n=N_1}^{N_2} \left| y(n) + \sum_{i=1}^{p} a(i)y(n-i) \right|^2 \tag{4.22}$$

From the complex number properties, we know that the absolute value square of a set of complex number is equal to the square values of the real and imaginary parts of each number (in our case only the real factors are present) and, therefore, the summing of these factors from N_1 to N_2 is equivalent to the Frobenius norm of the following matrix (see Appendix B):

$$J(\mathbf{a}) = \left\| \begin{bmatrix} y(N_1) \\ y(N_1+1) \\ \vdots \\ y(N_2) \end{bmatrix} + \begin{bmatrix} y(N_1-1) & \cdots & y(N_1-p) \\ y(N_1) & \cdots & y(N_1+1-p) \\ \vdots & & \vdots \\ y(N_2-1) & \cdots & y(N_2-p) \end{bmatrix} \begin{bmatrix} a(1) \\ a(1) \\ \vdots \\ a(p) \end{bmatrix} \right\|^2 = \|\mathbf{a} + \mathbf{Ya}\|^2 \tag{4.23}$$

where we assumed that $y(n) = 0$ for $n < N_1$ and $n > N_2$.

The vector $\mathbf{a}$ that minimizes the cost function $J(\mathbf{a})$ is given by

$$\boxed{\hat{\mathbf{a}} = -(\mathbf{Y}^T\mathbf{Y})^{-1}(\mathbf{Y}^T\mathbf{y})} \tag{4.24}$$

If we set $N_1 = p + 1$ and $N_2 = N$, the above vectors take the form:

$$\mathbf{y} = \begin{bmatrix} y(p+1) \\ y(p+2) \\ \vdots \\ y(N) \end{bmatrix}, \quad \mathbf{Y} = \begin{bmatrix} y(p) & y(p-1) & \cdots & y(1) \\ y(p+1) & y(p) & \cdots & y(2) \\ \vdots & \vdots & & \vdots \\ y(N-1) & y(N-2) & \cdots & y(N-p) \end{bmatrix} \tag{4.25}$$

Example 4.2

Use the following AR model to find 64 terms of its output:

$$y(n) = 1.3847y(n-1) - 1.5602y(n-2) + 0.8883y(n-3) - 0.4266y(n-4) + v(n)$$

The noise v is a zero mean and unit variance Gaussian process. In addition, the following signal was added:
$s(n) = 3\sin(0.3n\pi) + 3\sin(0.32n\pi) + 2\sin(0.6n\pi)$.

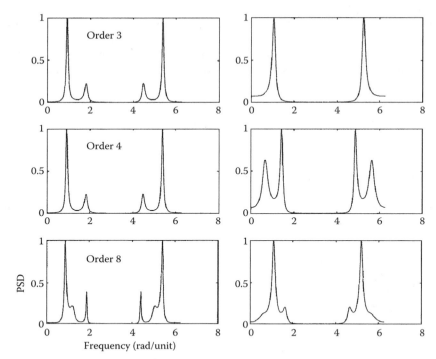

Figure 4.3

Solution

Figure 4.3 shows two columns. In the first column, the PSD for order 3, order 4 (equal to the AR model), and order 8 was produced with noise variance of 1. The second column was produced with variance of the noise of about 100. It is apparent from the figure that the order, as well as the signal-to-noise power, plays a predominant role in identifying line spectra.   ■

The following Book MATLAB function was used to create Figure 4.3. The reader can easily change the IIR filter producing the output and/or the signal.

Book MATLAB function: ssp_covariance_methodpsd(p,N)

```
function[y,a]=ssp_covariance_methodpsd(p,N)
%p=order of the AR filter for the PSD estimate;
%N=the number of IIIR output and signal length;
%y=signal and IIR output combined;a is a vector
```

```
%of the AR coefficients;
for m=0:N-1
y1(1)=0;y1(2)=0;y1(3)=0;y1(4)=0;
y1(m+5)=1.3847*y1(m+4)-1.5602*y1(m+3)+0.8883*y1(m+2)...
  -0.4266*y1(m+1)+randn;
s(m+1)=3*sin(0.3*pi*m)+3*sin(0.32*pi*m) +2*sin(0.6*pi*m);
end;
y=y1(1,5:length(y1))+s;
for t=0:-p+N-1
    yv(t+1)=y(t+p+1);
end;
for r=0:N-p-1
    for q=0:p-1
    Y(r+1,q+1)=y(1,p+r-q);
end;
end;
a=-(inv(Y'*Y))*(Y'*yv');
```

To produce the graph, at the command window you can write

```
w=0:2*pi/512:2*pi-(2*pi/512);
m=1:p;
psd=1./abs((1+exp(-j*w'*m)*a)).^2));
plot(w,psd);
```

Modified covariance method using linear transformation

To produce Figure 4.4, we used the following Book MATLAB function:

Book MATLAB function: [y,a,Y]=ssp_mod_cov_meth1(p,N)

```
function[y,a,Y]=ssp_mod_cov_meth1(p,N)
%N=length of data in 2^n;p=order of approximating AR
%system;
for m=0:N-1
    y1(1)=0;y1(2)=0;y1(3)=0;y1(4)=0;
    y1(m+5)=1.3847*y1(m+4)-1.5602*y1(m+3)...
        +0.8 883*y1(m+2)-.4266*y1(m+1) +randn;
  s(m+1)=3*sin(0.3*pi*m) +3*sin(0.33*pi*m) +2*sin(0.6*pi*m);
end;
y=y1(1,5:length(y1))+s;
ylt=[y*0.2 0.2*y+.05*randn(1,length(y))];
for t=0:-p+length(ylt)-1
    yv(t+1)=ylt(t+p+1);
end;
```

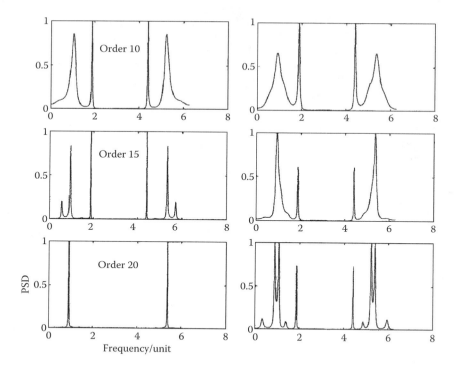

Figure 4.4

```
for r=0:length(ylt)-p-1
    for q=0:p-1
    Y(r+1,q+1)=ylt(1,p+r-q);
  end;
end;
a=-(inv(Y'*Y))*(Y'*yv');
```

On the left-hand side of Figure 4.4, we plotted the PSD produced by the covariance method versus the order of the AR estimating system. On the right-hand side, we plotted the PSD produced by the proposed modified method of linear transformation of random variables (rvs). It is apparent that the proposed method successfully differentiates the two frequencies at 0.3π and 0.33π units apart.

4.4   Minimum variance (MV) method

The MV method produces a spectrum estimate by filtering the signal through a bank of narrowband band-pass filters. Let us have a band-pass filter that has unit gain at ω_i. This constraint dictates that

$$H_i(e^{j\omega_i}) = \sum_{n=0}^{p-1} h_i(n)e^{j\omega_i n} = 1, \quad \mathbf{h}_i = [h_i(0) \quad h_i(1) \quad \cdots \quad h_i(p-1)]^T,$$

$$\mathbf{e}_i = [1 \quad e^{j\omega_i} \quad \cdots \quad e^{j\omega_i(p-1)}]^T \tag{4.26}$$

Therefore, Equation 4.26 becomes

$$\mathbf{h}_i^H \mathbf{e}_i = \mathbf{e}_i^H \mathbf{h}_i = 1 \tag{4.27}$$

conjugate each vector element and then transpose the vector. For real data, the superscript H becomes the transpose operator T. The spectrum of the output of the above band-pass filter (system) is given by Equation 2.45 to be equal to

$$E\{|y_i(n)|^2\} = \mathbf{h}_i^H \mathbf{R}_x \mathbf{h}_i \tag{4.28}$$

It can be shown that the filter design problem becomes one of minimizing Equation 4.28 subject to the linear constraint given by Equation 4.27. Hence, we write

$$\mathbf{h}_i = \frac{\mathbf{R}_x^{-1} \mathbf{e}_i}{\mathbf{e}_i^H \mathbf{R}_x^{-1} \mathbf{e}_i} \tag{4.29}$$

for one of the filters in the bank and for frequency ω_i. Because this is any frequency, the above equation is true for all frequencies. Hence, we write

$$\mathbf{h} = \frac{\mathbf{R}_x^{-1} \mathbf{e}}{\mathbf{e}^H \mathbf{R}_x^{-1} \mathbf{e}} \tag{4.30}$$

and the power estimate is

$$\hat{\sigma}_x^2(e^{j\omega}) \triangleq E\{|y(n)|^2\} = \mathbf{h}^H \mathbf{R}_x \mathbf{h} = \frac{\mathbf{e}^H \mathbf{R}_x^{-1}}{\mathbf{e}^H \mathbf{R}_x^{-1} \mathbf{e}} \mathbf{R}_x \frac{\mathbf{R}_x^{-1} \mathbf{e}}{\mathbf{e}^H \mathbf{R}_x^{-1} \mathbf{e}}$$

$$= \frac{\mathbf{e}^H \mathbf{R}_x^{-1} \mathbf{e}}{\mathbf{e}^H \mathbf{R}_x^{-1} \mathbf{e} \, \mathbf{e}^H \mathbf{R}_x^{-1} \mathbf{e}} = \frac{1}{\mathbf{e}^H \mathbf{R}_x^{-1} \mathbf{e}} \tag{4.31}$$

where $\mathbf{e} = [1 \ e^{j\omega} \cdots e^{j\omega(p-1)}]^T$. The reader should remember that $\mathbf{R}$ is symmetric and, therefore, $\mathbf{R} = \mathbf{R}^H$, $\mathbf{R}^{-H} = \mathbf{R}^{-1}$.

Figure 4.5 shows the PSD using the MV method and averaging 20 spectrums with increasing the correlation matrix from 5×5 to 20×20 in increments of 5. Figure 4.5a shows the signal $x(n) = 3\sin(0.3\pi n) + 3\sin(0.32\pi n) + 3\sin(0.6\pi n) + \text{rand}(1,256)$, and Figure 4.5b shows the PSD. The following Book MATLAB function produces the power estimate using the MV method.

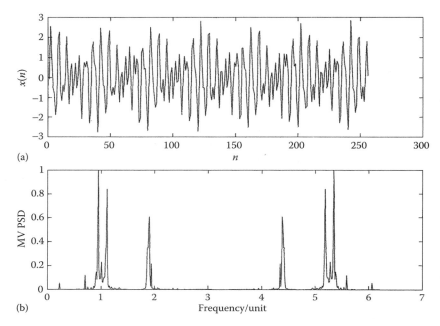

Figure 4.5

Book MATLAB function: [psd] = ssp_mv_psd1(x,P,L)

```
function[psd]=ssp_mv_psd1(x,P,L)
%x=data;P=number divided exactly by 5;
%L=number of frequency bins;
%psd=averaged power spectra P/5 times;
r=ssp_sample_biased_autoc(x,floor(length(x)*0.8));
psd1=0;
for p=5:5:P;
    R=toeplitz(r(1,1:p));
for m=0:L-1;
    n=0:p-1;
e=exp(-j*m*(2*pi/L)*(n-1));
ps(m+1) =abs(1/(e*inv(R)*(conj(e))'));
end;
psd1=psd1+abs(ps)/max(abs(ps));
end;
psd=psd1/max(psd1);
```

Example 4.3

Find the power spectrum estimate using the MV approach. Assume a white noise signal $\{x(n)\}$ having zero mean value and variance σ_x^2. The autocorrelation matrix is a 4×4 matrix.

Solution

The autocorrelation of the white noise signal is given by $r_x(n,m) = \sigma_x^2 \, \delta(n-m)$. Therefore, the autocorrelation matrix takes the form:

$$\mathbf{R}_x = \begin{bmatrix} \sigma_x^2 & 0 & 0 & 0 \\ 0 & \sigma_x^2 & 0 & 0 \\ 0 & 0 & \sigma_x^2 & 0 \\ 0 & 0 & 0 & \sigma_x^2 \end{bmatrix} = \sigma_x^2 \begin{bmatrix} 1 & 0 & 0 & 0 \\ 0 & 1 & 0 & 0 \\ 0 & 0 & 1 & 0 \\ 0 & 0 & 0 & 1 \end{bmatrix} = \sigma_x^2 \mathbf{I}$$

Since we have the following relations (see Appendix B for matrix inversion):

$$\mathbf{R}_x^{-1} = \frac{1}{\sigma_x^2} \mathbf{I}, \quad \mathbf{e}^H = \begin{bmatrix} 1 & e^{-j\omega} & e^{-j2\omega} & e^{-j3\omega} \end{bmatrix}, \quad \mathbf{e} = \begin{bmatrix} 1 & e^{j\omega} & e^{j2\omega} & e^{j3\omega} \end{bmatrix}^T$$

Equation 4.31 becomes

$$\hat{\sigma}_x^2(e^{j\omega}) \triangleq E\{|y(n)|^2\} = \frac{1}{\mathbf{e}^H \mathbf{R}_x^{-1} \mathbf{e}} = \frac{\sigma_x^2}{4} \qquad \blacksquare$$

4.5   Model order

It is important that we are able to approximately estimate the order of the system from which the data were produced. To estimate the order of the filter, Akaike has developed two criteria based on concepts in mathematical statistical. These are the **final prediction error** (FPE) and the **Akaike information criterion** (AIC). Their corresponding equations are

$$\text{FPE} = s_p^2 \frac{N+p}{N-p} \qquad (4.32a)$$

$$\text{AIC} = N \ln(s_p^2) + 2p \qquad (4.32b)$$

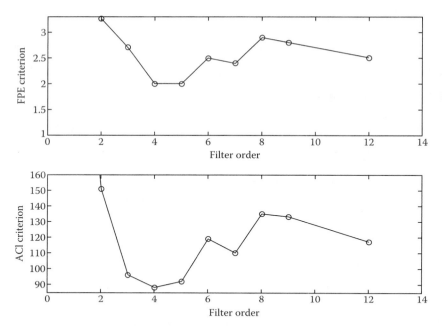

Figure 4.6

Studies on the FPE criterion show that it tends to have a minimum at values of the order p are less than the model order. The shortcomings of the AIC criterion, is that it tends to overestimate the model order. Besides these short-comings both of these criteria are often used in practical applications. Figure 4.6 shows both of these criteria using the sspAR_FPE_AIC_criterion1 m-file to produce the needed $a(i)$s coefficients and the autocorrelation function $r(k)$. It is obvious to conclude, from both criteria, that the order is about 4, which is the correct value. For the present case, we used 128 data points and 30 lag times for the autocorrelation function.

4.6   *Levinson–Durbin algorithm*

The Levinson–Durbin algorithm is a recursive algorithm for solving the YW equations to estimate the model coefficients. This scheme is based upon the concept of estimating the parameters of the model of order p from the parameters of a model of order $p - 1$. This is possible because of the Toeplitz form of the matrix (see Appendix B). From Equation 4.10, with $m = 0$ and $k = 1$, and taking into consideration that the impulse response is zero for negative times, we obtain (the as are estimates):

$$r_y(1) = -a_1(1)r_y(0) \quad \text{or} \quad a_1(1) = -\frac{r_y(1)}{r_y(0)} \qquad (4.33)$$

From Equation 4.12, and for **first-order** system, we find the relation:

$$\sigma_{v,1}^2 = r_y(0)(1 - a_1(1)^2)$$

(4.34)

After finding the above two initial relations, the recursive scheme starts from the second-order model and is found from Equation 4.14 to be equal to

$$\begin{bmatrix} r_y(0) & r_y(1) \\ r_y(1) & r_y(0) \end{bmatrix} \begin{bmatrix} a_2(1) \\ a_2(2) \end{bmatrix} = \begin{bmatrix} -r_y(1) \\ -r_y(2) \end{bmatrix}$$

(4.35)

Observe that we used the equality of the correlation coefficients $r_y(k) = r_y(-k)$ and specifically identified the order of approximation of the as coefficients with a subscript. Using Equation 4.35 and solving for $a_2(1)$, we obtain

$$r_y(0)a_2(1) = -r_y(1) - r_y(1)a_2(2) \quad \text{or} \quad a_2(1) = -\frac{r_y(1)}{r_y(0)} - a_2(2)\frac{r_y(1)}{r_y(0)}$$

$$= a_1(1) + a_2(2)a_1(1)$$

(4.36)

To obtain $a_2(2)$, we must use the augmented YW equation (Equation 4.13) of order 2:

$$\begin{bmatrix} r_y(0) & r_y(1) & r_y(2) \\ r_y(1) & r_y(0) & r_y(1) \\ r_y(p) & r_y(1) & r_y(0) \end{bmatrix} \begin{bmatrix} 1 \\ a_2(1) \\ a_2(2) \end{bmatrix} = \begin{bmatrix} \sigma_{v,2}^2 \\ 0 \\ 0 \end{bmatrix}$$

(4.37)

The crucial point is to express the left-hand side of Equation 4.37 in terms of $a_2(2)$. Using Equation 4.36, the column vector of as becomes

$$\begin{bmatrix} 1 \\ a_2(1) \\ a_2(2) \end{bmatrix} = \begin{bmatrix} 1 \\ a_1(1) + a_1(1)a_2(2) \\ a_2(2) \end{bmatrix} = \begin{bmatrix} 1 \\ a_1(1) \\ 0 \end{bmatrix} + a_2(2) \begin{bmatrix} 0 \\ a_1(1) \\ 1 \end{bmatrix}$$

(4.38)

Introducing the results of the above equation in Equation 4.37 and applying the matrix properties (see Appendix B), we obtain

$$\begin{bmatrix} r_y(0) & r_y(1) & r_y(2) \\ r_y(1) & r_y(0) & r_y(1) \\ r_y(2) & r_y(1) & r_y(0) \end{bmatrix} \begin{bmatrix} 1 \\ a_1(1) \\ 0 \end{bmatrix} + a_2(2) \begin{bmatrix} r_y(0) & r_y(1) & r_y(2) \\ r_y(1) & r_y(0) & r_y(1) \\ r_y(2) & r_y(1) & r_y(0) \end{bmatrix} \begin{bmatrix} 1 \\ a_1(1) \\ 0 \end{bmatrix}$$

$$= \begin{bmatrix} r_y(0) + r_y(1)a_1(1) \\ r_y(1) + r_y(0)a_1(1) \\ r_y(2) + r_y(1)a_1(1) \end{bmatrix} + a_2(2) \begin{bmatrix} r_y(1)a_1(1) + r_y(2) \\ r_y(0)a_1(1) + r_y(1) \\ r_y(1)a_1(1) + r_y(0) \end{bmatrix} = \begin{bmatrix} \sigma_{v,2}^2 \\ 0 \\ 0 \end{bmatrix}$$

(4.39)

The above equation can be written in the form:

$$
\mathbf{t}_1 + \mathbf{t}_2 =
\begin{bmatrix}
r_y(0)[1 - a_1(1)^2] \\
r_y(1) + r_y(0)a_1(1) \\
r_y(2) + r_y(1)a_1(1)
\end{bmatrix}
+ a_2(2)
\begin{bmatrix}
r_y(1)a_1(1) + r_y(2) \\
r_y(0)a_1(1) + r_y(1) \\
r_y(0)[1 - a_1(1)^2]
\end{bmatrix}
$$

$$
=
\begin{bmatrix}
\sigma_{v,1}^2 \\
\Delta_2 \\
\Delta_3
\end{bmatrix}
+ a_2(2)
\begin{bmatrix}
\Delta_3 \\
\Delta_2 \\
\sigma_{v,1}^2
\end{bmatrix}
=
\begin{bmatrix}
\sigma_{v,2}^2 \\
0 \\
0
\end{bmatrix}
\tag{4.40}
$$

From the third equation of the above system, we obtain the unknown $a_2(2)$ as follows:

$$
\Delta_3 + a_2(2)\sigma_{v,1}^2 = 0 \quad \text{or} \quad a_2(2) = -\frac{\Delta_3}{\sigma_{v,1}^2} = -\frac{r_y(1)a_1(1) + r_y(2)}{\sigma_{v,1}^2}
\tag{4.41}
$$

Using Equation 4.41 and the first equation of the system (Equation 4.40), we obtain the variance for step two or first order:

$$
\sigma_{v,2}^2 = \sigma_{v,1}^2 + a_2(2)\Delta_3 = (1 - a_2(2)^2)\sigma_{v,1}^2
\tag{4.42}
$$

Example 4.4

The autocorrelation matrix of a second-order IIR system output, with an input of zero mean Gaussian noise, is given by

$$
R =
\begin{bmatrix}
5.5611 & 1.3291 & 0.8333 \\
1.3291 & 5.5611 & 1.3291 \\
0.8333 & 1.3291 & 5.5611
\end{bmatrix}
$$

Proceed to find the variances and filter coefficients.

Solution

From $a_1(1) = -[r_y(1)/r_y(0)] = -(1.3291/5.5611) = -0.2390$ and from Equation 4.34, we obtain $\sigma_{v,1}^2 = r_y(0)(1 - a_1(1)^2) = 5.5611(1 - 0.2390^2) = 5.2434$. From Equation 4.41, we obtain

$$
a_2(2) = -\frac{\Delta_3}{\sigma_{v,1}^2} = -\frac{r_y(1)a_1(1) + r_y(2)}{\sigma_{v,1}^2} = -\frac{-(5.5611 \times 0.2390) + 1.3291}{5.2434} = -0.1929
$$

and from Equation 4.36, we obtain

$$a_2(1) = a_1(1) + a_2(2)a_1(1) = -0.2390 - 0.1929(-0.2390) =$$
-0.1929. The noise variance is $\sigma_{v,2}^2 = \sigma_{v,1}^2 + a_2(2)\Delta_3 = (1 - a_2(2)^2)\sigma_{v,1}^2 = (1 - (-0.1929)^2)5.2434 = 5.0483$. ■

Levinson–Durbin algorithm (real-valued case)

1. Initialize the recursion with the zero-order model
   a. $a_0(0) = 1$
   b. $\sigma_{v,0}^2 = r_y(0)$
2. For $j = 0, 1, 2, \ldots, p - 1$
   a. $k_j = r_y(j+1) + \sum_{i=1}^{j} a_j(i) r_y(j - i + 1)$
   b. $K_{j+1} = -k_j / \sigma_{v,j}^2$
   c. For $i = 1, 2, \ldots, j$
      $$a_{j+1}(i) = a_j(i) + K_{j+1} a_j(j - i + 1)$$
   d. $a_{j+1}(j + 1) = K_{j+1}$
   e. $\sigma_{v,j+1}^2 = \sigma_{v,j}^2 \left[1 - K_{j+1}^2\right]$
3. $b(0) = \sqrt{\sigma_{v,p}^2}$

Book MATLAB function: [a,var] = ssp_levinson(r,p)

```
function [a,var]=ssp_levinson(r,p)
r=r(:);%p=number of a's+1;to find r we can use
        %the Book MATLAB function
        %r=sasamplebiasedautoc(x,lg), p<lg,
        %lg=the lag number of the autocorelation
        %function r;
a=1;
var=r(1);
for j=2:p
    g=r(2:j)'*flipud(a);
    gamma=-g/var;
    a=[a;0]+gamma*[0;(flipud(a))];
    var=var*(1-abs(gamma)^2);
end;
```

Based on the above results, we can proceed to plot the spectrum using the following expression in the command window:

```
ps=abs(sqrt(var)./(exp(-j*[0:0.01:6.2]'*[1:p])*[a])).^2);
plot([0:0.01:6.2],ps/max(ps));
```

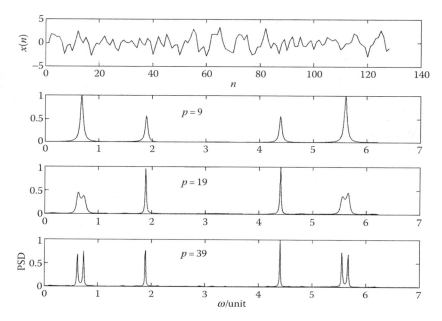

Figure 4.7

Figure 4.7 shows the original signal and the spectrum for three different number (10, 20, and 40) of ps which correspond to as = $p - 1$ coefficients of the AR model. The signal was $x = \sin(0.2\pi n) + \sin(0.25\pi n) + \sin(0.6\pi n) + 0.5*\mathrm{randn}(1, 128)$.   ∎

Similarly, we can also use the YW function below to obtain the desired coefficients.

Book MATLAB Yule–Walker function:
[a,var] = ssp_yule_walker_alg (x,lg,p)

```
function[a,var]=ssp_yule_walker_alg(x,lg,p)
%x=signal;lg=lag number;
%p=order of AR model=number of a's<lg;
r=sasamplebiasedautoc(x,lg);
R=toeplitz(r(1,1:p));
a=inv(R)*r(1,2:p+1)';
var=r(1)+sum(r(1,p+1).*a');
```

To plot the spectrum we must write in the command window:

```
>>ps=abs(sqrt(var)./(exp(-j*[0:0.01:6.2]'*[1:length(a)...
+1])*[1;-a]).^2);
>>plot([0:0.01:6.2],ps/max(ps));
```

4.7 Maximum entropy method

Since we are limited to small number of lags when finding the autocorrelation function, it is natural to ask the question: how can we expand the autocorrelation function such that the extension be as accurate as possible. One such method that was suggested is the **maximum entropy method** (MEM). It can be shown that the estimate of the PSD using the MEM approach is given by

$$\boxed{S_{\text{MEM}}(e^{j\omega}) = \frac{b(0)^2}{|\mathbf{e}^H\mathbf{a}|^2}}$$

(4.43)

where

$$\begin{bmatrix} r_x(0) & r_x(-1) & \cdots & r_x(-p) \\ r_x(1) & r_x(0) & \cdots & r_x(p-1) \\ \vdots & \vdots & & \vdots \\ r_x(p) & r_x(p-1) & \cdots & r_x(0) \end{bmatrix} \begin{bmatrix} 1 \\ a(1) \\ \vdots \\ a(p) \end{bmatrix} = \begin{bmatrix} b(0)^2 \\ 0 \\ \vdots \\ 0 \end{bmatrix}$$

(4.44)

$$b(0)^2 = r_x(0) + \sum_{k=1}^{p} a(k)r_x(k)$$

(4.45)

$$\mathbf{e}^H = \begin{bmatrix} 1 & e^{-j\omega} & e^{-j\omega^2} & \cdots & e^{-j\omega p} \end{bmatrix}$$

(4.46)

The vector **a** is the solution of Equation 4.44 and $r_x(k)$ is the correlation of the data. First, the autocorrelation normal equations (Equation 4.44) are solved for the as and $b(0)^2$. The following Book MATLAB function produces the PSD using the MEM.

Book MATLAB function: function[s] = sspmax_entropy_meth(x,p)

```
function[s]=ssp_max_entropy_meth(x,lg,p)
%x=data;p=order of AR system;p<lg=lag number of the
%autocorrelation function;
[a,var]=ssp_yule_walker_alg(x,lg,p);
s=var./abs(fft([1;-a],512));%s=PSD;
w=0:2*pi/512:2*pi-(2*pi/512);
```

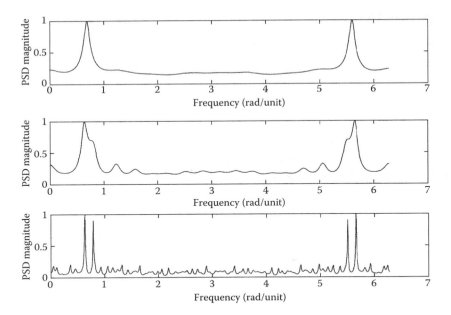

Figure 4.8

To plot the spectrum, we write in the command window:

```
>>plot(w,s/max(s));
```

Figure 4.8 shows the PSD using the MEM with the following inputs: $x(n) = \sin(0.2\pi n) + \sin(0.23\pi n) + \sin(0.6\pi n) + 0.5*\text{randn}(1,128)$, lg = 60, and $p = 10, 20,$ and 40.

4.8   Spectrums of segmented signals

Many times we are faced with the situation where some data are missing, e.g., breaking down of a detector or receiving data from space and interrupted by the rotation of the earth.

Let the sequence to be used is

$$x(n) = \sin(0.2\pi n) + \sin(0.25\pi n) + v(n) \tag{4.47}$$

Let the sequence $\{x(n)\}$ has 64 elements of which $x(17) = x(18) = \cdots = x(48) = 0$ (are missing). The noise $v(n)$ is WG of zero mean and variance σ_v^2. Figure 4.9 shows the sequence, its segmented form, and their corresponding spectra.

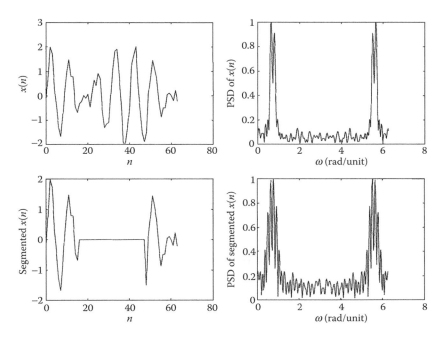

Figure 4.9

Method 1: The average method

In this method, we obtain the spectrum, S_1, of the first segment, next the spectrum, S_2, of the second segment and then average them: $S = (S_1 + S_2)/2$. Figure 4.10a shows the spectrum of the first segment, Figure 4.10b shows the spectrum of the second nonzero segment, and Figure 4.10c shows the average spectrum of the first two spectra. The signal is the same as given in the previous paragraph.

Method 2: Extrapolation method

One of the basic ideas is to assume that a signal is the result of an AR process and find extra values of the signal using the extrapolation method. Let the estimate $x(n)$ be given by the linear combination of the previous values of the system. Hence,

$$\hat{x}(n) = -a_1 x(n-1) - a_2 x(n-2) - \cdots - a_p x(n-p) \tag{4.48}$$

then, the error in the estimate is given by the difference

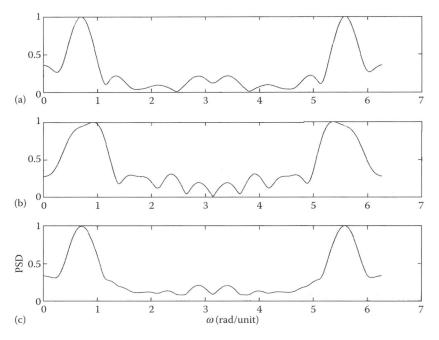

Figure 4.10

$$\varepsilon(n) = x(n) - \hat{x}(n) \tag{4.49}$$

or

$$\varepsilon(n) = x(n) + a_1 x(n-1) + a_2 x(n-2) + \cdots + a_p x(n-p) \tag{4.50}$$

The normal equations, whose solution provides the optimal coefficients, can be derived directly in matrix form by first defining the vectors:

$$x(n) = \begin{bmatrix} x(n-p) \\ x(n-p+1) \\ \vdots \\ x(n) \end{bmatrix}, \quad a = \begin{bmatrix} 1 \\ a_1 \\ \vdots \\ a_p \end{bmatrix}, \quad \tilde{x}(n) = \begin{bmatrix} x(n) \\ x(n-1) \\ \vdots \\ x(n-p) \end{bmatrix}, \quad l = \begin{bmatrix} 1 \\ 0 \\ \vdots \\ 1 \end{bmatrix} \tag{4.51}$$

The vector a, of the coefficients and the prediction error variance $\sigma_\varepsilon^2 = E\{\varepsilon^2(n)\}$ constitute the **linear prediction parameters**. Based on Equation 4.51, we write (see Equation 4.50):

$$\varepsilon(n) = a^T \tilde{x}(n) \tag{4.52}$$

To find the optimal filter coefficients, we apply the orthogonality principle, which states that $E\{x(n-i)\varepsilon(n)\} = 0$, $i = 1, 2, \ldots, p$ and $\sigma_\varepsilon^2 = E\{x(n)\varepsilon(n)\}$. These assertions can be stated in compact matrix form:

$$E\{\tilde{x}(n)\varepsilon(n)\} = \begin{bmatrix} \sigma_\varepsilon^2 \\ 0 \\ 0 \\ \vdots \\ 0 \end{bmatrix} = \sigma_\varepsilon^2 l \tag{4.53}$$

Drop the argument n for simplicity and substitute Equation 4.52 in Equation 4.53 to obtain

$$E\{\tilde{x}(a^T \tilde{x})\} = E\{\tilde{x}(\tilde{x}^T a)\} = E\{\tilde{x}\tilde{x}^T\}a = \sigma_\varepsilon^2 l \quad \text{or} \quad \boxed{Ra = \sigma_\varepsilon^2 l} \tag{4.54}$$

Equation 4.54 is the well-known normal equations.

Example 4.5

> The following Book MATLAB function finds the spectrum of a segmented sequence with 64 elements of which the 32 central ones are missing. The results are shown in Figures 4.11. The signal $x(n) = \sin(0.3\pi n) + \sin(0.35\pi n) + 0.5\,\text{randn}(1,64)$ with 64 values is segmented as follows: $xs(n) = [x(1, 1:18)\ \text{zeros}(1, 28)\ x(1, 47:64)]$. Figure 4.11a presents the noisy signal, Figure 4.11b shows the segmented signal, and Figure 4.11c shows the PSD of the noisy signal. Figure 4.11d shows the PSD of the segmented signal, Figure 4.11e shows the PSD of the forward extension using the first segment, and Figure 4.11f shows the PSD of the backward extension of the second segment. It is apparent that the extrapolation gives good results.

Book MATLAB function:
[x1,a1,xe1,x2,a2,xe2,xe1f,xe2f]=ssp_psd_meth_1(x,seg1,seg2,p]

```
function[x1,a1,xe1,x2,a2,xe2,xe1f,xe2f]=...
ssp_extrap_psd_meth_1(x,seg1,seg2,p)
%r=autocorrelation of x;seg1=length of first
%segment;seg2=length
```

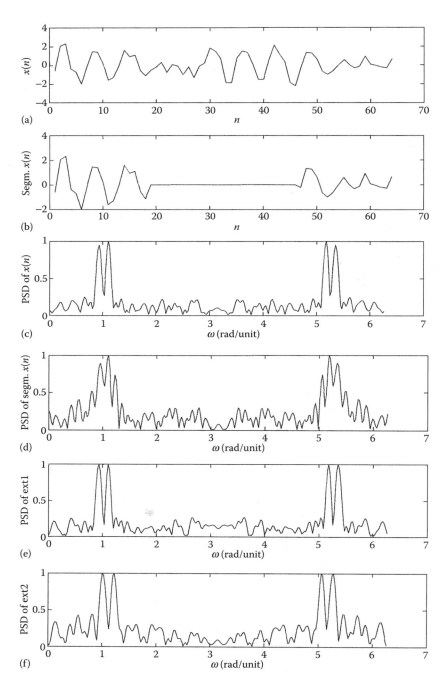

Figure 4.11

```
%of the second segment (se fig
%2);length(x)=2^n;p=filter order;
subplot(3,1,1);plot(x,'k');xlabel('n');ylabel('x(n)');
xx12=[x(1,1:seg1) zeros(1,(length(x)-seg1-seg2))...
  x(1,(length(x)-seg2+1):length(x))];
subplot(3,1,2);plot(xx12,'k');xlabel('n');ylabel('Segm....
  x(n)');. .
%--------------
%Spectrum of the complete sample and segment samples;
w=0:2*pi/256:2*pi-(2*pi/256);
fx=fft(x,256);
subplot(3,1,3);plot(w,abs(fx)/max(abs(fx)),'k');
xlabel('\omega rad/unit');ylabel('PSD of x(n)');
fxs=fft(xx12,256);
figure(2);
subplot(3,1,1);plot(w,abs(fxs)/max(abs(fxs)),'k');
xlabel('\omega rad/unit');ylabel('PSD of segm. x(n)');
%--------------
%Extrapolation
r1=xcorr(x(1,1:seg1),'biased');
rs1=r1(1,seg1:2*seg1-1);%p<length(rs1);
[a1,var1]=ssp_yule_walker_alg_rp(rs1,p);
Ext1=length(x)-seg1-seg2;
for m=1:Ext1
    x1(seg1+m)=sum(a1'.*fliplr(x(1,seg1-p+m:seg1-1+m)));
end;
xe1=[x(1,1:length(x)-seg1-seg2) x1(length(x)-seg1-...
  seg2+1:length(x)...
  -seg2) x(1,length(x)-seg2+1:length(x))];
r2=xcorr(x(1,length(x)-seg2:length(x)));
rs2=r2(1,seg2:2*seg2-1);
[a2,var2]=ssp_yule_walker_alg_rp(rs2,p);
%the above p can change to p1 if different number of a's
%is desired; p<min{rs1,rs2};
Ext2=length(x)-seg1-seg2;
for k=1:Ext2
    x2(length(x)-seg2+1-k)=-sum ((a2/max(a2))'....
      *x(1,length(x)-seg2+2-k:length(x)-seg2+1+p-k));
end;
xe2=[x(1,1:length(x)-seg1-seg2) x2(length(x)...
  -seg1-seg2+1:length(x)
  -seg2) x(1,length(x)-seg2+1:length(x))];
xe1f=(abs(fft(xe1,256))/max(abs(fft(xe1,256))));
subplot(3,1,2)
```

```
plot(w,xe1f,'k');xlabel('\omega rad/unit');ylabel ...
  ('PSD of ext1');
subplot(3,1,3);
xe2f=(abs(fft(xe2,256))/max(abs(fft(xe2,256))));
plot(w,xe2f,'k');xlabel('\omega rad/unit');ylabel ...
  ('PSD ext2');
```                                                                              ■

We can further introduce the following approaches for the extrapolation method: (1) from the forward extrapolated signal, find the autocorrelation function and then the PSD; (2) from the backward extrapolated signal, find the autocorrelation signal and then its PSD; (3) find the average of the previous two spectrums; and (4) using Monte Carlo method by repeating the process and then averaging.

4.9   Eigenvalues and eigenvectors of matrices (see also Appendix B)

Let us create a relationship between a vector v and a $n \times n$ matrix $\mathbf{A}$ and a constant λ as follows:

$$\mathbf{A}\mathbf{q} = \lambda\mathbf{q} \quad \text{or} \quad (\mathbf{A} - \lambda\mathbf{q}) = 0 \tag{4.55}$$

where $\mathbf{I}$ is the identity matrix having 1 along the diagonal and 0 for the rest of the elements. Since $\mathbf{q} = 0$ is always a solution, we must find the non-zero solution if it exists. The number λ is called the **eigenvalue** and the vector $\mathbf{q}$ belonging to that particular eigenvalue is called **eigenvector**. To find the eigenvalues, the determinant $|\mathbf{A} - \lambda\mathbf{I}|$ must be equal to 0. If, for example,

$$\mathbf{A} = \begin{bmatrix} 4 & -5 \\ 2 & -3 \end{bmatrix}, \quad \text{then} \ |\mathbf{A} - \lambda\mathbf{I}| = \begin{Vmatrix} 4 & -5 \\ 2 & -3 \end{Vmatrix} - \begin{bmatrix} \lambda & 0 \\ 0 & \lambda \end{bmatrix} = \begin{vmatrix} 4-\lambda & -5 \\ 2 & -3-\lambda \end{vmatrix} = 0$$

then

$$(4-\lambda)(-3-\lambda) + 10 = 0 \quad \text{or} \quad \lambda^2 - \lambda - 2 = 0 \quad \text{or} \quad \lambda_1 = -1, \ \lambda_2 = 2$$

MATLAB has the following function to find the eigenvalues and the corresponding eigenvectors: **[Q,D]** = eig(A), where **Q** is a matrix containing the eigenvectors (columns) and **D** is a diagonal matrix containing the eigenvalues. For this case, and referring to Equation 4.55, we have the relations:

$$\lambda_1 = -1, \quad (\mathbf{A} - \lambda\mathbf{I})\mathbf{q} = \begin{bmatrix} 5 & -5 \\ 2 & -2 \end{bmatrix}\begin{bmatrix} q_1 \\ q_2 \end{bmatrix} = \begin{bmatrix} 0 \\ 0 \end{bmatrix}, \quad \begin{cases} 5q_1 - 5q_2 = 0 \\ 2q_1 - 2q_2 = 0 \end{cases}$$

It is apparent that the solution to the system is any eigenvector which is the multiple of the vector $\mathbf{q}_1 = [1 \quad 1]^T$. Similarly, the second eigenvalue corresponds to any multiple of the vector $\mathbf{q}_2 = [5 \quad 2]^T$. If we use the MATLAB function, we obtain

$$[\mathbf{Q},\mathbf{D}] = \mathrm{eig}(\mathbf{A}), \quad \mathbf{Q} = \begin{bmatrix} 0.9285 & 0.7071 \\ 0.3714 & 0.7071 \end{bmatrix}, \quad \mathbf{D} = \begin{bmatrix} 2 & 0 \\ 0 & -1 \end{bmatrix}$$

In the above equation, the columns of $\mathbf{Q}$ are the eigenvectors and the diagonal values are the corresponding eigenvalues. Although the eigenvector matrix, given by MATLAB, seems that gives different results, the values are proportional. If we divide the first column by 0.3714 and multiply by 2, we obtain the vector $[5 \quad 2]^T$. Similarly, if we divide the second by 0.7071, we obtain the vector $[1 \quad 1]^T$.

After finding the eigenvectors and eigenvalues, the following matrix operations are true

$$\mathbf{AQ} = \mathbf{QD} \quad \text{or} \quad \mathbf{Q}^{-1}\mathbf{AQ} = \mathbf{D} \quad \text{or} \quad \mathbf{A} = \mathbf{Q}^{-1}\mathbf{DQ} \qquad (4.56)$$

It is known (see Problem 4.9.1) in matrix theory that

> *If the eigenvectors of a matrix correspond to different eigenvalues, then those eigenvectors are linear independent. Therefore, if the eigenvectors of a matrix are linearly independent then the eigenvalues are distinct. If, in addition, the matrix is symmetric, the eigenvectors corresponding to distinct eigenvalues are orthogonal.*

Eigendecomposition of the autocorrelation matrix

Let us consider a sinusoid having random phase

$$x(n) = A\sin(n\omega_0 + \phi) \qquad (4.57)$$

where A and ω_0 are fixed constants and the phase is a rv that is uniformly distributed over the interval $-\pi \le \phi \le \pi$. The mean value and the autocorrelation of $x(n)$ is given by (see Problem 4.9.2)

$$E\{x(n)\} \triangleq m_x(n) = 0, \quad E\{x(k)x(l)\} \triangleq r_x(k,l) = \frac{1}{2}A^2\cos[(k-l)\omega_0] \qquad (4.58)$$

Since the mean is a constant and the autocorrelation depends only on the difference (lag time) the process is wide-sense stationary (WSS).

If a signal is made up of sinusoids with additive white noise

$$x(n) = \sum_{m=1}^{M} A_m \sin(n\omega_m + \phi_m) + v(n) \qquad (4.59)$$

its autocorrelation function is

$$r_x(k,l) = \frac{1}{2} \sum_{m=1}^{M} A_m^2 \cos[(k-l)\omega_m] + r_v(k,l) \tag{4.60}$$

where the amplitudes and frequencies are fixed constants and the phases are rvs that are uniformly distributed.

We can consider the complex form of the sinusoids as follows:

$$x(n) = A e^{j(n\omega_0 + \phi)}$$

where the phase is a rv uniformly distributed in the interval $-\pi \le \phi \le \pi$. Therefore, the mean is (expand the exponent in Euler's format):

$$m_x(n) = E\{A e^{j(n\omega_0 + \phi)}\} = 0 \tag{4.61}$$

and the autocorrelation function is

$$\begin{aligned} r_x(k,l) &= E\{x(k)x^*(l)\} = E\{A e^{j(n\omega_0 + \phi)} A^* e^{-j(n\omega_0 + \phi)}\} \\ &= |A|^2 E\{e^{j(k-l)\omega_0}\} = P e^{j(k-l)\omega_0} = r_x(k-l) \end{aligned} \tag{4.62}$$

where $k - l$ is the lag of the autocorrelation function and can be substituted simply by k (setting $l = 0$). P is the power of the sinusoidal (complex) signal. Note that the mean is constant and the autocorrelation is independent of the time origin. Therefore, the process is a WSS.

If the noise is white with mean value zero, then the autocorrelation (see Section 2.2) function is $\sigma_v^2 \delta(n)$ and, therefore, the autocorrelation of p sinusoids embedded in noise is

$$r_x(k) = \sum_{i=1}^{p} P_i e^{j\omega_i k} + \sigma_v^2 \delta(k) \tag{4.63}$$

For $p = 2$, we obtain

$$r_x(0) = P_1 + P_2 + \sigma_v^2, \quad \delta(0); \qquad r_x(1) = P_2 e^{j\omega_1} + P_2 e^{j\omega_2}, \quad \delta(1) = 0$$

Therefore, the correlation matrix is

$$\mathbf{R} = \begin{bmatrix} r_x(0) & r_x(1) \\ r_x(1) & r_x(0) \end{bmatrix}$$

If we define (the exponent H stands for Hermitian or, equivalently, stands for transposed conjugate of a vector or matrix and just conjugation for a complex quantity)

$$\mathbf{e}_1 = [1 \quad e^{j\omega_1}]^T, \quad \mathbf{e}_1^H = [1 \quad e^{-j\omega_1}], \quad \mathbf{e}_2 = [1 \quad e^{j\omega_2}]^T, \quad \mathbf{e}_2^H = [1 \quad e^{-j\omega_2}]$$

then (see Appendix B)

$$\mathbf{R}_x = P_1 \mathbf{e}_1 \mathbf{e}_1^H + P_2 \mathbf{e}_2 \mathbf{e}_2^H + \sigma_v^2 \mathbf{I} = \mathbf{R}_s + \mathbf{R}_n$$

The above equation can be put in the compact form:

$$\mathbf{R}_x = \mathbf{E}\mathbf{P}\mathbf{E}^H + \sigma_v^2 \mathbf{I}, \quad \mathbf{E} = [\mathbf{e}_1 \quad \mathbf{e}_2], \quad \mathbf{E}^H = [\mathbf{e}_1^H \quad \mathbf{e}_2^H]^T, \quad \mathbf{P} = \begin{bmatrix} P_1 & 0 \\ 0 & P_2 \end{bmatrix} \quad (4.64)$$

Therefore, for p sinusoids, the general autocorrelation matrix is

$$\mathbf{R}_x = \mathbf{R}_s + \mathbf{R}_n = \sum_{i=1}^{p} P_i \mathbf{e}_i \mathbf{e}_i^H + \sigma_v^2 \mathbf{I},$$

$$\mathbf{e}_i = [1 \quad e^{j\omega_i} \quad e^{j2\omega_i} \quad \cdots \quad e^{j(M-1)\omega_i}], \quad i = 1, 2, \ldots, p \quad (4.65)$$

Example 4.6

Find the eigenvalues and eigenvectors of a signal having the following correlation matrices:

$$\mathbf{R}_s = \begin{bmatrix} 0.5 & 0.25 \\ 0.25 & 1.00 \end{bmatrix}, \quad \mathbf{R}_v = \begin{bmatrix} 0.1 & 0 \\ 0 & 0.1 \end{bmatrix}, \quad \mathbf{R}_x = \mathbf{R}_s + \mathbf{R}_v = \begin{bmatrix} 0.6 & 0.25 \\ 0.25 & 1.10 \end{bmatrix}$$

Solution

Using MATLAB, we obtain

1. Signal eigenvectors and eigenvalues

$$\mathbf{Q}_s = [\mathbf{q}_1 \quad \mathbf{q}_2] = \begin{bmatrix} -0.9239 & 0.3827 \\ 0.3827 & -0.9239 \end{bmatrix}, \quad \mathbf{D}_s = \begin{bmatrix} 0.3964 & 0 \\ 0 & 1.1036 \end{bmatrix},$$

$$\mathbf{q}_1^H * \mathbf{q}_1 = 1, \quad \mathbf{q}_2^H * \mathbf{q}_2 = 1, \quad \mathbf{q}_1^H * \mathbf{q}_2 = 0$$

2. Noise eigenvectors and eigenvalues

$$\mathbf{Q}_v = [\mathbf{q}_1 \quad \mathbf{q}_2] = \begin{bmatrix} 1 & 0 \\ 0 & 1 \end{bmatrix}, \quad \mathbf{D}_v = \begin{bmatrix} 0.1 & 0 \\ 0 & 0.1 \end{bmatrix}$$

$$\mathbf{q}_1^H * \mathbf{q}_1 = 1, \quad \mathbf{q}_2^H * \mathbf{q}_2 = 1, \quad \mathbf{q}_1^H * \mathbf{q}_2 = 0$$

3. Data eigenvectors and eigenvalues

$$\mathbf{R}_x \mathbf{v}_i = (\mathbf{R}_s + \sigma_v^2 \mathbf{I}) \mathbf{v}_i = \lambda_i^s \mathbf{v}_i + \sigma_v^i \mathbf{v}_i = (\lambda_i^s + \sigma_v^i) \mathbf{v}_i \quad (4.66)$$

Therefore, the eigenvectors of $\mathbf{R}_x$ are the same as those of $\mathbf{R}_s$, and the eigenvalues of $\mathbf{R}_x$ are

$$\lambda_i = \lambda_i^s + \sigma_v^i$$

Therefore, the largest eigenvalue of $\mathbf{R}_x$ is

$$\lambda_{max} = MP_1 + \sigma_v^2$$

and the remaining $M - 1$ eigenvalues are equal to σ_v^2.

Note: Parameter extraction for one frequency in the data

1. Perform an eigendecomposition of the autocorrelation matrix $\mathbf{R}_x$. The largest eigenvalue is equal to $\lambda_{max} = MP_1 + \sigma_v^2$ and the remaining eigenvalues are equal to σ_v^2.
2. Use the eigenvalues of $\mathbf{R}_x$ to solve for the power P_1 and the noise variance

$$\sigma_v^2 = \lambda_{min}, \quad P_1 = \frac{1}{M}(\lambda_{max} - \lambda_{min})$$

3. Since $\mathbf{R}_x$ is the result of noisy data $\{x(n)\}$, we consider weighted averages as follows. Let $\mathbf{v}_i$ be a noise eigenvector of $\mathbf{R}_x$, e.g., one that has an eigenvalue σ_v^2, and let $v_i(k)$ be the kth component of $\mathbf{v}_i$. If we compute the DTFT, of the coefficients in $\mathbf{v}_i$

$$V_i(e^{j\omega}) = \sum_{k=0}^{M-1} v_i(k)e^{-jk\omega} = \mathbf{e}^H \mathbf{v}_i \tag{4.67}$$

then the orthogonality condition (see Equation 4.64) implies that at $\omega = \omega_i$, the value of $V_i(e^{j\omega})$ will be equal to 0 and, hence, the **PSD estimation function**

$$\hat{S}(e^{j\omega}) = \frac{1}{\left|\sum_{k=0}^{M-1} v_i(k)e^{-jk\omega}\right|^2} = \frac{1}{\left|\mathbf{e}^H \mathbf{v}_i\right|^2} \tag{4.68}$$

will be extremely large at $\omega = \omega_i$. This is an effective way to estimate the frequency. To avoid errors in estimating the frequency, while using only one eigenvector, it is recommended that a weighted average of all the noise eigenvectors are used. Hence, we write

$$\hat{S}(e^{j\omega}) = \frac{1}{\sum_{k=1}^{M-1} a_i \left|\mathbf{e}^H \mathbf{v}_i\right|^2} \tag{4.69}$$

where a_is are some appropriate chosen constants.

Harmonic model

Consider the signal model that consists of p complex exponentials in white noise:

$$x(n) = \sum_{i=1}^{p} A_i e^{j\omega_i n} + v(n) \tag{4.70}$$

where ω_i is the discrete-time frequency (rad/unit), and A_i is a complex number of the form

$$A_i = |A_i| e^{j\phi_i} \tag{4.71}$$

It is known (see Chapters 1 and 2) that the spectrum of sinusoidal functions are impulses in the frequency domain. The additive noise produces a constant background level at the power level of the white noise

$$\sigma_v^2 = E\{|v(n)|^2\}.$$

Since we will be dealing with discrete signals, it is advantageous to form a vector of the signal over a time window of length M. Therefore, the data take the form:

$$\mathbf{x}(n) = [x(n) \quad x(n+1) \quad x(n+2) \quad \cdots \quad x(n+M-1)]^{T} \tag{4.72}$$

The signal in Equation 4.70 can know be written in the form:

$$\mathbf{x}(n) = \sum_{i=1}^{p} A_i e(\omega_i) e^{j\omega_i n} + \mathbf{v}(n) = \mathbf{s}(n) + \mathbf{v}(n)$$

$$\mathbf{x}(n) = \sum_{i=1}^{p} A_i e(\omega_i) e^{j\omega_i n} = \text{signal}, \quad \mathbf{v}(n) = [v(n) \quad v(n+1) \quad \cdots \quad v(n+M-1)]^{T}$$

$$e(\omega_i) = [1 \quad e^{j\omega_i} \quad e^{j\omega_i 2} \quad \cdots \quad e^{j\omega_i(M-1)}]^{T} = \text{time window frequency vector} \tag{4.73}$$

Note that we differentiate the signal made up of sinusoidal functions (complex) and the white noise signal. We observe that for the data $x(n) = s(n) + v(n)$, the autocorrelation function of $x(n)$ is

$$r_x(k) = r_s(k) + \sigma_v^2 \delta(k), \quad k = 0, \pm 1, \pm 2, \ldots, \pm(M-1) \tag{4.74}$$

Therefore, the $M \times M$ autocorrelation matrix of $x(n)$ may be expressed as

$$\mathbf{R}_x = \mathbf{R}_s + \mathbf{R}_v = \sum_{i=1}^{p} |A_i|^2 \, e(\omega_i) e^{H}(\omega_i) + \sigma_v^2 \, \mathbf{I} = \mathbf{E}\mathbf{A}\mathbf{E}^{H} + \sigma_v^2 \mathbf{I}$$

$$\mathbf{E} = [e(\omega_1) \quad e(\omega_2) \quad \cdots \quad e(\omega_p)] \tag{4.75}$$

where $\mathbf{E}$ is a $M \times p$ matrix whose columns are the time-window frequency vectors of length M (see Equation 4.73) at frequencies ω_i and $\mathbf{A}$ stands for the matrix

$$\mathbf{A} = \begin{bmatrix} |A_1|^2 & 0 & \cdots & 0 \\ 0 & |A_2|^2 & \cdots & 0 \\ \vdots & \vdots & & \vdots \\ 0 & 0 & \cdots & |A_p|^2 \end{bmatrix} \tag{4.76}$$

We must always take the time-window length M to be greater than the number of sinusoids p.

We can write the $M \times M$ autocorrelation matrix of the data in the form (see Appendix B):

$$\mathbf{R}_x = \mathbf{Q}\Lambda\mathbf{Q}^H = \sum_{i=1}^{M} \lambda_i \mathbf{q}_i \mathbf{q}_i^H, \quad \mathbf{Q} = [\mathbf{q}_1 \quad \mathbf{q}_2 \quad \cdots \quad \mathbf{q}_M], \quad \Lambda = \begin{bmatrix} \lambda_1 & 0 & \cdots & 0 \\ 0 & \lambda_2 & \cdots & 0 \\ \vdots & \vdots & \vdots & \vdots \\ 0 & 0 & \cdots & \lambda_M \end{bmatrix} \tag{4.77}$$

Correspondingly, we can also write the above equation as follows:

$$\mathbf{R}_x = \sum_{i=1}^{p} \lambda_i \mathbf{q}_i \mathbf{q}_i^H + \sum_{i=p+1}^{M} \sigma_v^2 \mathbf{q}_i \mathbf{q}_i^H = \mathbf{Q}_s \Lambda_s \mathbf{Q}_s^H + \sigma_v^2 \mathbf{Q}_v \mathbf{Q}_v^H,$$

$$\mathbf{Q}_s = [\mathbf{q}_1 \quad \mathbf{q}_2 \quad \cdots \quad \mathbf{q}_p], \quad \mathbf{Q}_v = [\mathbf{q}_{p+1} \quad \mathbf{q}_{p+2} \quad \cdots \quad \mathbf{q}_M] \tag{4.78}$$

Thus, the M-dimensional subspace that contains the observations of the time-window signal vector can be split into two subspaces spanned by the signal and noise eigenvectors, respectively. These two subspaces are known as the **signal subspace** and **noise subspace**. These subspaces are orthogonal to each other since the correlation matrix is Hermitian symmetric (the eigenvectors of a Hermitian symmetric matrix are orthogonal $\mathbf{q}_j^H \mathbf{q}_i$).

The eigendecomposition separates the eigenvectors into two sets. The set $\{\mathbf{q}_1 \mathbf{q}_2 \cdots \mathbf{q}_p\}$ that is principal eigenvectors, span the signal subspace. The set $\{\mathbf{q}_{p+1} \mathbf{q}_{p+2} \cdots \mathbf{q}_M\}$ that is orthogonal to the principal eigenvectors belongs to the noise subspace. Since the signal vectors $\mathbf{e}_i$ are in the signal subspace, it simply follows that they are a linear combination of the principal eigenvectors and, hence, they are orthogonal to the vectors in the noise subspace (see Figure 4.12).

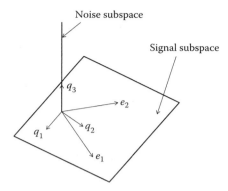

Figure 4.12

Example 4.7

Find the eigenvalues and eigenvectors of a signal having the following correlation matrices:

$$\mathbf{R}_s = \begin{bmatrix} 0.5 & 0.25 \\ 0.25 & 1.0 \end{bmatrix}, \quad \mathbf{R}_v = \begin{bmatrix} 0.1 & 0 \\ 0 & 0.1 \end{bmatrix}, \quad \mathbf{R}_x = \mathbf{R}_s + \mathbf{R}_v = \begin{bmatrix} 0.6 & 0.25 \\ 0.25 & 1.1 \end{bmatrix}$$

Solution

Using MATLAB, we obtain

1. Signal eigenvectors and eigenvalues

$$\mathbf{Q}_s = [\mathbf{q}_1 \quad \mathbf{q}_2] = \begin{bmatrix} -0.9239 & 0.3827 \\ 0.3827 & 0.9239 \end{bmatrix}, \quad \mathbf{D}_s = \begin{bmatrix} 0.3964 & 0 \\ 0 & 1.1036 \end{bmatrix}$$

$$\mathbf{q}_1^H * \mathbf{q}_1 = 1, \quad \mathbf{q}_2^H * \mathbf{q}_2 = 1, \quad \mathbf{q}_1^H * \mathbf{q}_2 = 0$$

2. Noise eigenvectors and eigenvalues

$$\mathbf{Q}_v = [\mathbf{q}_1 \quad \mathbf{q}_2] = \begin{bmatrix} 1 & 0 \\ 0 & 1 \end{bmatrix}, \quad \mathbf{D}_v = \begin{bmatrix} 0.1 & 0 \\ 0 & 0.1 \end{bmatrix}$$

$$\mathbf{q}_1^H * \mathbf{q}_1 = 1, \quad \mathbf{q}_2^H * \mathbf{q}_2 = 1, \quad \mathbf{q}_1^H * \mathbf{q}_2 = 0$$

3. Data eigenvector and eigenvalues

$$\mathbf{Q}_x = [\mathbf{q}_1 \quad \mathbf{q}_2] = \begin{bmatrix} -0.9239 & 0.3827 \\ 0.3827 & 0.9239 \end{bmatrix}, \quad \mathbf{D}_x = \begin{bmatrix} 0.4964 & 0 \\ 0 & 1.2036 \end{bmatrix}$$

$$\mathbf{q}_1^H * \mathbf{q}_1 = 1, \quad \mathbf{q}_2^H * \mathbf{q}_2 = 1, \quad \mathbf{q}_1^H * \mathbf{q}_2 = 0$$

4. Expansion of autocorrelation matrix (the exponent H (Hermitian) becomes T (transpose) for real matrices)

$$\mathbf{Q}_s \mathbf{D}_s \mathbf{Q}_s^H + \sigma_v^2 \mathbf{Q}_v \mathbf{Q}_v^H = \begin{bmatrix} -0.9239 & 0.3827 \\ 0.3827 & 0.9239 \end{bmatrix} \begin{bmatrix} 0.3964 & 0 \\ 0 & 1.1036 \end{bmatrix} \begin{bmatrix} -0.9239 & 0.3827 \\ 0.3827 & 0.9239 \end{bmatrix}$$

$$+ 0.1 \begin{bmatrix} 1 & 0 \\ 0 & 1 \end{bmatrix} \begin{bmatrix} 1 & 0 \\ 0 & 1 \end{bmatrix} = \begin{bmatrix} 0.6000 & 0.2500 \\ 0.2500 & 1.1001 \end{bmatrix}$$

Based on the above results we conclude, as before, that the following relation holds:

$$\boxed{\mathbf{R}_x = \mathbf{Q}_s \mathbf{D}_s \mathbf{Q}_s^H + \sigma_v^2 \mathbf{Q}_v \mathbf{Q}_v^H, \quad \mathbf{Q}_s = [\mathbf{q}_1 \quad \mathbf{q}_2 \quad \cdots \quad \mathbf{q}_p], \quad \mathbf{Q}_v = [\mathbf{q}_{p+1} \quad \cdots \quad \mathbf{q}_M]} \quad (4.79)$$

Note: *(1) The eigenvalues of the data (signal plus noise) is equal to the sum of the eigenvalues of the signal plus the variance of the noise; (2) the eigenvectors of the data are identical to the eigenvectors of the signal; (3) Equation 4.79 relation holds.* ∎

If there exist *p* sinusoids in the data, then the first *p* eigenvalues in descending order correspond to the first part of Equation 4.79 (signal) and the remaining correspond to the second part (noise). These columns of these matrices consist of the signal and noise eigenvectors. This first part of Equation 4.79 is the signal subspace and the second part is the noise subspace.

The following book MATLAB function produces the autocorrelation matrix if the data vector (sequence) {*x(n)*} is given. The autocorrelation is given by

$$\hat{\mathbf{R}}_x = \frac{1}{N} \mathbf{X}^H \mathbf{X}, \quad \mathbf{X} = \begin{bmatrix} \mathbf{x}^T(1) \\ \mathbf{x}^T(2) \\ \vdots \\ \mathbf{x}^T(n) \\ \vdots \\ \mathbf{x}^T(N-1) \\ \mathbf{x}^T(N) \end{bmatrix} = \begin{bmatrix} x(1) & x(2) & \cdots & x(M) \\ x(2) & x(3) & \cdots & x(M+1) \\ \vdots & \vdots & & \vdots \\ x(n) & x(n+1) & \cdots & x(n+M-1) \\ \vdots & \vdots & & \vdots \\ x(N-1) & x(N) & \cdots & x(N+M-2) \\ x(N) & x(N+1) & \cdots & x(N+M-1) \end{bmatrix} \quad (4.80)$$

Book MATLAB function: [rx] = ssp_est_autocor_matrix(x)

```
function [rx] =ssp_est_autocor_matrix(x,M)
%rx=NxM matrix;N/M=integer;N=length(x)=2^n;
%M=time window along the vector x;x=row data;
N=length(x);
for n=1:N
    for m=1:M
    X(n,m)=x(1,m+n-1);
  end;
end;
rx=(conj(X'))*X/length(x);
```

Pisarenko harmonic decomposition

Pisarenko observed that an ARMA process consisting of p sinusoids in additive white noise, the noise variance corresponds to the minimum eigenvalue of R_x and, hence, the method proceeds as follows: (1) estimate R_x from the data; (2) find the minimum eigenvalue (the MATLAB function $[Q,D] = \text{eig}(R_x)$ gives the eigenvalues in the descent order and to those values correspond the eigenvectors); (3) find the minimum eigenvector; and (4) use the equation

$$\boxed{S(e^{j\omega}) = \frac{1}{|q^H e|^2}}$$

(4.81)

Example 4.8

> Find the spectrum (known as pseudospectrum) using the method of Pisarenko harmonic decomposition. The following data were used for Figure 4.13: $x(n) = \sin(0.5*\pi*n) + a*\text{randn}(1,32)$ with $n = 0, 1, 2, \ldots, 31$ and $a = 3$ and 1 which controls the variance of the noise. The following Book MATLAB function was used to obtain the spectrum:

Book MATLAB function for Pisarenko harmonic decomposition: [qmin,sigma,q,d]=ssp_pisar_harm_decomp(x,p)

```
function [qmin,sigma,q,d]=ssp_pisar_harm_decomp(x,p)
%this function graphs the PSD using Pisarenko
%harmonic decomposition; qmin=minimum eigenvector;
%sigma=noise variance;q=matrix eigenvectors;
%d=diagonal matrix of the eigenvalues by descending
%order;p=number of harmonic present;
```

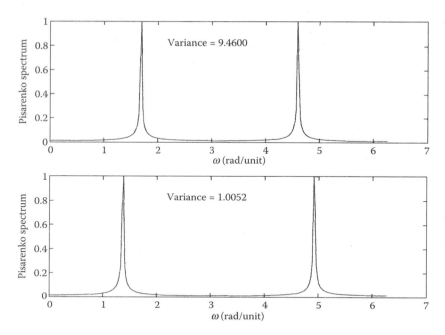

Figure 4.13

```
r=xcorr(x,'biased');
R=toeplitz(r(:,length(x):length(x)+2*p));
[q,d]=eig(R);
sigma=min(diag(d));
coln=find(diag(d)==sigma);
qmin=q(:,coln);
w=0:2*pi/256:2*pi-(2*pi/256);
plot(w,((abs(1./fft(qmin,256)))/abs...
   (1./fft(qmin,256)))),'k');
```

We observe that the strength of the noise plays a dominant effect on the location of the spectra lines. The Pisarenko method is important from the conceptual and analytical perspective. It lucks robustness to be used for most applications. Furthermore, the correlation matrix must be estimated and, therefore, the resulting noise eigenvectors are only estimated. Because the roots of the minimum estimator can be close to the unit circle, splitting of the line spectrum can also occur.

MUSIC algorithm

The **multiple signal classification** (MUSIC) algorithm is based on two disciplines: (1) the window M is not set equal to $p + 1$ but larger than that, and (2) the average of the noise spectra. Therefore, we write

$$\left| S_{music}(e^{j\omega}) = \frac{1}{\sum_{i=1}^{M-p} |e^H q_i|^2} = \frac{1}{\sum_{i=1}^{M-p} |Q_i(e^{j\omega})|^2} \right.,$$

$$e = [1 \quad e^{j\omega} \quad e^{j\omega 2} \quad \cdots \quad e^{j\omega(M-1)}], Q_i(e^{j\omega}) = \text{FT of } i\text{th eigenvecto} \qquad (4.82)$$

It has also been suggested to multiply the ith factor of the summation by $1/\lambda_i$. The following Book MATLAB function plots the MUSIC spectrum.

Book MATLAB function for the MUSIC spectrum:
[psd,q,d,R] = ssp _music_alg(x,p,M,nfft)

```
function[psd,q,d,R]=ssp_music_alg(x,p,M,nfft)
%length(x)+M<2*length(x); R=correlation matrix;
%p=number of sinusoids; M=time window;nfft=the
%desired number of bins in fft e.g. 256 or 512;
%to observe the spectrum write plot(psd/max(psd));
r=xcorr(x,'biased');
R=toeplitz(r(:,length(x):length(x)+M));
[q,d]=eig(R);
qin=zeros(nfft,1);
for i=1:M-p-1
    qin=qin+abs(fft(q(:,i),nfft));
end;
psd=1./(qin/max(qin));
```

Figure 4.14 shows the MUSIC spectrum for the following function and for two time windows: (a) $M = 5$ and (b) $M = 15$,

$$x(n) = \sin(0.3^*\pi^*n) + \sin(0.6^*\pi^*n) + \text{rand}(1, 32), \quad n = 0, 1, 2, \ldots, 31.$$

Problems

4.1.1   Derive Equation 4.10.
4.3.1   Verify Equation 4.19.
4.3.2   Verify Equation 4.20.
4.3.3   Verify Equation 4.21.

Solutions, suggestions, and hints

4.1.1   From Equation 4.1, we write

$$y(n) + \sum_{m=1}^{p} a(m)y(n-m) = \sum_{m=0}^{q} b(m)v(n-m), \qquad b(0) = 1 \qquad (1)$$

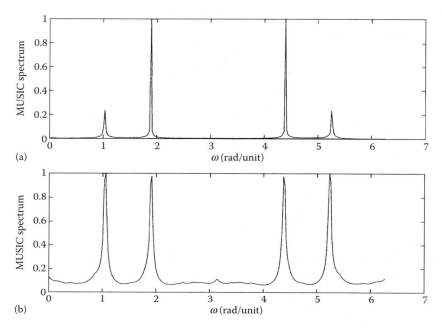

(a)

(b)

Figure 4.14

Multiply (1) by $y(n - k)$ and take expectations to obtain

$$E\{y(n)y(n-k)\} + \sum_{m=1}^{p} a(m)E\{y(n-m)y(n-k)\} = \sum_{m=0}^{q} b(m)E\{v(n-m)y(n-k)\}$$

or

$$r_y(k) + \sum_{m=1}^{p} a(m)r_y(k-m) = \sum_{m=0}^{q} b(m)E\{v(n-m)y(n-k)\} \qquad (2)$$

but

$$y(n) = \sum_{s=0}^{\infty} h(s)v(n-s) \qquad (3)$$

and, hence the last expectation of (2) becomes

$$E\{v(n-m)y(n-k)\} = E\left\{v(n-m)\sum_{s=0}^{\infty} h(s)v(n-k-s)\right\}$$

$$= \sum_{s=0}^{\infty} h(s)E\{v(n-m)v(n-k-s)\}$$

$$= \sum_{s=0}^{\infty} h(s)\sigma_v^2\delta(k+s-m) = \sigma_v^2 h(m-k)$$

that implies (2) becomes

$$r_y(k) + \sum_{m=1}^{p} a(m) r_y(k-m) = \sigma_v^2 \sum_{m=0}^{q} b(m) h(m-k)$$

4.3.1 $\sigma_p^2 \triangleq E\{|e(n)|^2\} = E\{[y(n) + \mathbf{a}^T \mathbf{y}(n)][y(n) + \mathbf{y}^T(n)\mathbf{a}]\}$

$$= E\{y^2(n)\} + \mathbf{a}^T E\{\mathbf{y}(n) y(n)\} + E\{y(n) \mathbf{y}^T(n)\}\mathbf{a} + \mathbf{a}^T E\{\mathbf{y}(n)\mathbf{y}^T(n)\}\mathbf{a}$$

$$r(0) = E\{y^2(n)\} ; \quad \mathbf{a}^T \mathbf{r}_{y(p)} = \mathbf{a}^T E\{\mathbf{y}(n) y(n)\} ; \quad \mathbf{r}_{y(p)}^T \mathbf{a} = E\{y(n)\mathbf{y}^T(n)\}\mathbf{a};$$

$$\mathbf{a}^T \mathbf{R}_{y(p)} \mathbf{a} = \mathbf{a}^T E\{\mathbf{y}(n)\mathbf{y}^T(n)\} \mathbf{a}$$

4.3.2   From Equation 4.19, we obtain (for a 2×2 matrices, which can be extrapolated to any dimension [see also Appendix B]):

$$\frac{\partial \sigma^2}{\partial a(1)} = [r(1) \quad r(2)] \begin{bmatrix} 1 \\ 0 \end{bmatrix} + [1 \quad 0] \begin{bmatrix} r(1) \\ r(2) \end{bmatrix} + [1 \quad 0] \begin{bmatrix} r(1) & r(2) \\ r(2) & r(1) \end{bmatrix} \begin{bmatrix} a(1) \\ a(2) \end{bmatrix}$$

$$+ [a(1) \quad a(2)] \begin{bmatrix} r(1) & r(2) \\ r(2) & r(1) \end{bmatrix} \begin{bmatrix} 1 \\ 0 \end{bmatrix} = 2r(1) + 2r(1)a(1) + 2r(2)a(2) = 0,$$

$$r(1) + [r(1) \quad r(2)] \begin{bmatrix} a(1) \\ a(2) \end{bmatrix} = 0 \qquad (1)$$

$$\frac{\partial \sigma^2}{\partial a(2)} = [r(1) \quad r(2)] \begin{bmatrix} 0 \\ 1 \end{bmatrix} + [0 \quad 1] \begin{bmatrix} r(1) \\ r(2) \end{bmatrix} + [0 \quad 1] \begin{bmatrix} r(1) & r(2) \\ r(2) & r(1) \end{bmatrix} \begin{bmatrix} a(1) \\ a(2) \end{bmatrix}$$

$$+ [a(1) \quad a(2)] \begin{bmatrix} r(1) & r(2) \\ r(2) & r(1) \end{bmatrix} \begin{bmatrix} 0 \\ 1 \end{bmatrix} = 2r(2) + 2r(2)a(1) + 2r(1)a(2) = 0,$$

$$r(1) + [r(2) \quad r(1)] \begin{bmatrix} a(1) \\ a(2) \end{bmatrix} = 0 \qquad (2)$$

(1) and (2) above form the following matrix form:

$$\begin{bmatrix} r(1)a(1) + r(2)a(2) \\ r(2)a(1) + r(1)a(2) \end{bmatrix} = - \begin{bmatrix} r(1) \\ r(2) \end{bmatrix} \quad \text{or} \quad \mathbf{Ra} = -\mathbf{r}$$

4.3.3 $J(\mathbf{a}) = r_y(0) + \mathbf{r}_{y(p)}^T \mathbf{a} + \mathbf{a}^T \mathbf{r}_{y(p)} + \mathbf{a}^T \mathbf{R}_{y(p)} \mathbf{a}$ \qquad (1)

let $\mathbf{a} = \mathbf{a}_0 + \tilde{\mathbf{a}}$, $\mathbf{a}_0 = - \mathbf{R}_{y(p)}^{-1} \mathbf{r}_{y(p)}$ then (1) becomes

$$J(\mathbf{a}) = r_y(0) + \mathbf{r}_{y(p)}^T \mathbf{a} + \mathbf{a}^T \mathbf{r}_{y(p)} + \mathbf{a}^T \mathbf{R}_{y(p)} \mathbf{a}$$

$$= r_y(0) + \mathbf{r}_{y(p)}^T (\mathbf{a}_0 + \tilde{\mathbf{a}}) + (\mathbf{a}_0^T + \tilde{\mathbf{a}}^T) \mathbf{r}_{y(p)} + (\mathbf{a}_0^T + \tilde{\mathbf{a}}^T) \mathbf{R}_{y(p)} (\mathbf{a}_0 + \tilde{\mathbf{a}})$$

$$= r_y(0) - \mathbf{r}_{y(p)}^T \mathbf{R}_{y(p)}^{-1} \mathbf{r}_{y(p)} + \mathbf{r}_{y(p)}^T \tilde{\mathbf{a}} - \mathbf{r}_{y(p)}^T \mathbf{R}_{y(p)}^{-1} \mathbf{r}_{y(p)} + \tilde{\mathbf{a}}^T \mathbf{r}_{y(p)} \mathbf{r}_{y(p)}^T \mathbf{R}_{y(p)}^{-1} \mathbf{r}_{y(p)}$$

$$- \tilde{\mathbf{a}}^T \mathbf{r}_{y(p)} - \mathbf{r}_{y(p)}^T \tilde{\mathbf{a}} + \tilde{\mathbf{a}}^T \mathbf{R}_{y(p)} \tilde{\mathbf{a}}$$

$$= \tilde{\mathbf{a}}^T \mathbf{R}_{y(p)} \tilde{\mathbf{a}} + y(0) - \mathbf{r}_{y(p)}^T \mathbf{R}_{y(p)}^{-1} \mathbf{r}_{y(p)}$$

where we used the following identities (see also Appendix B):

$$\mathbf{R}_{y(p)}^{-1}\mathbf{R}_{y(p)} = \mathbf{I}, \qquad \tilde{\mathbf{a}}^T\mathbf{r}_{y(p)} = \mathbf{r}_{y(p)}^T\tilde{\mathbf{a}}$$

because $J(\mathbf{a})$ is a constant, the first factor after the last equality is constant since matrix $\mathbf{R}$ is positive definite matrix and, hence, the last two terms is the minimum mean square error.

chapter five

Optimal filtering—Wiener filters

5.1 Mean square error (MSE)

In this chapter, we develop a class of linear optimum discrete-time filters known as the **Wiener filters**. These filters are optimum in the sense of minimizing an appropriate function of the error, known as the **cost function**. The cost function that is commonly used in filter design optimization is the MSE. Minimizing MSE involves only second-order statistics (correlations), and leads to a theory of linear filtering that is useful in many practical applications. This approach is common to all optimum filter designs. Figure 5.1 shows the block diagram of the optimum filter problem.

The basic idea is to recover a desired signal $d(n)$ given a noisy observation $x(n) = d(n) + v(n)$, where both $d(n)$ and $v(n)$ are assumed to be wide-sense stationary (WSS) processes. Therefore, the problem can be stated as follows.

"Design a filter that produces an estimate $\hat{d}(n)$ using a linear combination of the data $x(n)$ such that the MSE function (cost function):

$$J = E\{(d(n) - \hat{d}(n))^2\} = E\{e^2(n)\} \tag{5.1}$$

is minimized."

Depending on how the data $x(n)$ and the desired signal $d(n)$ are related, there are four basic problems which need solution. These are filtering, smoothing, prediction, and deconvolution.

5.2 Finite impulse response (FIR) Wiener filter

Let the sample response (filter coefficients) of the desired filter be denoted by **w**. This filter will process the real-valued stationary process $\{x(n)\}$ to produce an estimate $\hat{d}(n)$ of the desired real-valued signal $d(n)$. Without loss of generality, we will assume, unless otherwise stated, that the processes $\{x(n)\}$, $\{d(n)\}$, etc. have zero mean values. Furthermore, assuming that the filter coefficients do

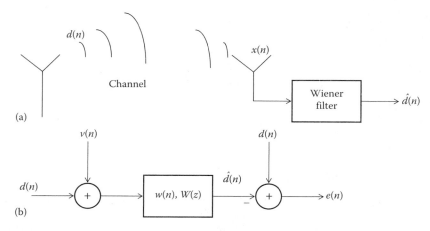

Figure 5.1

not change with time, the output of the filter is equal to the convolution of the input and the filter coefficients. Hence, we obtain

$$\hat{d}(n) = w(n) * x(n) = \sum_{m=0}^{M-1} w_m x(n-m) = \mathbf{w}^T \mathbf{x}(n) = \mathbf{x}^T(n)\mathbf{w} \tag{5.2}$$

where

$$\mathbf{w} = [w_0 \quad w_1 \quad w_2 \quad \cdots \quad w_{M-1}]^T$$
$$\mathbf{x}(n) = [x(n) \quad x(n-1) \quad \cdots \quad x(n-M)]^T$$

The MSE is given by (see Equation 5.1):

$$
\begin{aligned}
J(\mathbf{w}) &= E\{e^2(n)\} = E\{[d(n) - \mathbf{w}^T\mathbf{x}(n)]^2\} \\
&= E\{[d(n) - \mathbf{w}^T\mathbf{x}(n)][d(n) - \mathbf{w}^T\mathbf{x}(n)]^T\} \\
&= E\{d^2(n) - \mathbf{w}^T\mathbf{x}(n)d(n) - d(n)\mathbf{x}^T(n)\mathbf{w} + \mathbf{w}^T\mathbf{x}(n)\mathbf{x}^T(n)\mathbf{w}\} \\
&= E\{d^2(n)\} - 2\mathbf{w}^T E\{d(n)\mathbf{x}(n)\} + \mathbf{w}^T E\{\mathbf{x}(n)\mathbf{x}^T(n)\}\mathbf{w} \\
&= \sigma_d^2 - 2\mathbf{w}^T\mathbf{p}_{dx} + \mathbf{w}^T\mathbf{R}_x\mathbf{w}
\end{aligned}
\tag{5.3}
$$

where

$\mathbf{w}^T\mathbf{x}^T = \mathbf{x}^T(n)\mathbf{w} = $ number
$\sigma_d^2 = $ variance of the desired signal, $d(n)$
$\mathbf{p}_{dx} = [p_{dx}(0) \quad p_{dx}(1) \quad \cdots \quad p_{dx}(M-1)]^T = $ cross-correlation vector

$$p_{dx}(0) \overset{\Delta}{=} r_{dx}(0),\ p_{dx}(1) \overset{\Delta}{=} r_{dx}(1),\ \cdots,\ p_{dx}(M-1) \overset{\Delta}{=} r_{dx}(M-1) \tag{5.4}$$

$$\mathbf{R}_x = E\left\{\begin{bmatrix} x(n) \\ x(n-1) \\ \vdots \\ x(n-M+1) \end{bmatrix} \begin{bmatrix} x(n) & x(n-1) & \cdots & x(n-M+1) \end{bmatrix}\right\}$$

$$= \begin{bmatrix} E\{x(n)x(n)\} & E\{x(n)x(n-1)\} & \cdots & E\{x(n)x(n-M+1)\} \\ E\{x(n-1)x(n)\} & E\{x(n-1)x(n-1)\} & \cdots & E\{x(n-1)x(n-M+1)\} \\ \vdots & \vdots & \vdots & \vdots \\ E\{x(n-M+1)x(n)\} & E\{x(n-M+1)x(n-1)\} & \cdots & E(x(n-M+1)x(n-M+1)\} \end{bmatrix}$$

$$= \begin{bmatrix} r_x(0) & r_x(1) & \cdots & r_x(M-1) \\ r_x(-1) & r_x(0) & \cdots & r_x(M-2) \\ \vdots & \vdots & \vdots & \vdots \\ r_x(-M+1) & r_x(-M+2) & \cdots & r_x(0) \end{bmatrix} \tag{5.5}$$

The above matrix is the correlation matrix of the input data. The matrix is symmetric because the random process is assumed to be stationary and, hence, we have the equality, $r_x(k) = r_x(-k)$. Since in practical cases we have only one realization, we will assume that the signal is ergodic. Therefore, we will use the sample autocorrelation function, which is an estimate. However, in this chapter, we will not differentiate the estimate with an over bar since all of our simulations will be based only on estimate quantities and not on ensemble averages.

Example 5.1

Let us assume that we have found the sample autocorrelation coefficients $(r_x(0) = 1.0, r_x(1) = 0)$ from given data $x(n)$ which, in addition to noise, contain the desired signal. Furthermore, let the variance of the desired signal $\sigma_d^2 = 24$ and the cross-correlation vector be $\mathbf{p}_{dx} = [2 \ 4.5]^\mathrm{T}$. It is desired to find the surface defined by the mean-square function, $J(\mathbf{w})$.

Solution

Introducing the above values in Equation 5.3, we obtain

$$J(\mathbf{w}) = 24 - 2[w_0 \quad w_1]\begin{bmatrix} 2 \\ 4.2 \end{bmatrix} + [w_0 \quad w_1]\begin{bmatrix} 1 & 0 \\ 0 & 1 \end{bmatrix}\begin{bmatrix} w_0 \\ w_1 \end{bmatrix}$$

$$= 24 - 4w_0 - 9w_1 + w_0^2 + w_1^2$$

Note that the equation is quadratic with respect to filter coefficients and it is true for any number of filter coefficients. In the general case, products $w_i\, w_j$ will also be present. The data are the sum of the desired signal

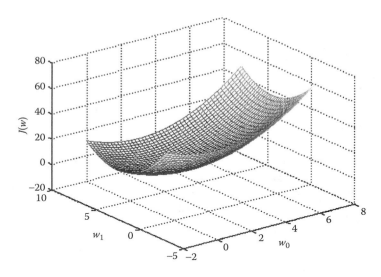

Figure 5.2

and noise. From the data we find the correlation matrix and the cross-correlation between the desired signal and the data. Note that to find the optimum Wiener filter coefficients the desired signal is needed. Figure 5.2 shows the **MSE surface**. This surface is found by inserting different values of w_0 and w_1 in the function $J(\mathbf{w})$. The values of the coefficients which correspond to the bottom of the surface are the **optimum** Wiener coefficients. The vertical distance from the w_0–w_1 plane to the bottom of the surface is known as the **minimum error**, $J_{\min}$, and corresponds to the optimum Wiener coefficients. We observe that the minimum height of the surface corresponds to about $w_0 = 2$ and $w_1 = 4.5$ which are the optimum coefficients as we will learn how to find them in the next section. The following book MATLAB® m-file was used:

Book MATLAB m-file: fig5_2_1

```
%Book MATLAB m-file:fig5_2_1

w0=-1:0.2:8;
w1=w0;
[W0,W1]=meshgrid(w0,w1);
%W0 and W1 are a pair of matrices
```

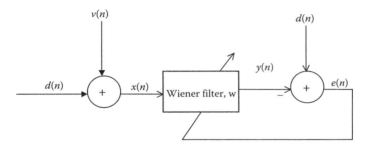

Figure 5.3

```
%representing a rectangular grid;
j=24-4*W0-9*W1+W0.^2+W1.^2;
colormap(gray)
mesh(w0,w1,j)
```

Figure 5.3 shows a schematic representation of the Wiener filter, and Figure 5.4 shows an adaptive FIR filter.

5.3   Wiener solution—orthogonal principle

From Figure 5.2, we observe that there exists a plane touching the parabolic surface at its minimum point, and is parallel to the w-plane. Furthermore, we observe that the surface is concave upward and, therefore from calculus, the first derivative of the MSE with respect to w_0 and w_1 must be zero at the minimum point and the second derivative must be positive. Hence, we write

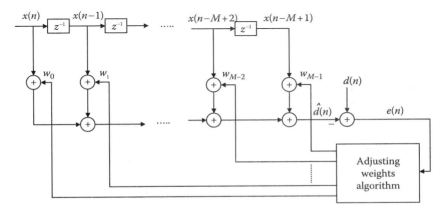

Figure 5.4

$$\frac{\partial J(w_0, w_1)}{\partial w_0} = 0, \quad \frac{\partial J(w_0, w_1)}{\partial w_1} = 0 \tag{5.6a}$$

$$\frac{\partial^2 (w_0, w_1)}{\partial w_0^2} > 0, \quad \frac{\partial^2 J(w_0, w_1)}{\partial w_1^2} > 0 \tag{5.6b}$$

For the two-coefficient filter, Equation 5.3 becomes

$$J(w_0, w_1) = w_0^2 r_x(0) + 2w_0 w_1 r_x(1) + w_1^2 r_x(0) - 2w_0 r_{dx}(0) - 2w_1 r_{dx}(1) + \sigma_d^2 \tag{5.7}$$

Introducing Equation 5.7 in Equation 5.6a produces the following set of equations:

$$2w_0^o r_x(0) + 2w_1^o r_x(1) - 2r_{dx}(0) = 0 \tag{5.8a}$$

$$2w_1^o r_x(0) + 2w_0^o r_x(1) - 2r_{dx}(1) = 0 \tag{5.8b}$$

The above system can be written in a form, called the discrete form of the Wiener–Hopf equation, which is

$$\boxed{\mathbf{R}_x \mathbf{w}^o = \mathbf{p}_{dx}} \tag{5.9}$$

The superscript "o" indicates the optimum Wiener solution for the filter. Note that to find the correlation matrix we must know the second-order statistics (autocorrelation). If, in addition, the matrix is invertible, the optimum filter is given by

$$\boxed{\mathbf{w}^o = \mathbf{R}_x^{-1} \mathbf{p}_{dx}} \tag{5.10}$$

It turns out, that in most practical signal processing applications, the matrix $\mathbf{R}_x$ is invertible. For an M-order filter, $\mathbf{R}_x$ is an $M \times M$ matrix, $\mathbf{w}^o$ is an $M \times 1$ vector, and $\mathbf{p}$ is an $M \times 1$ vector.

If we differentiate once again Equation 5.8 with respect to w_0^o and w_1^o, which is equivalent to differentiating $J(\mathbf{w})$ twice, we find that it is equal to $2r_x(0)$. But $r_x(0) = E\{x(m)x(m)\} = \sigma_x^2 > 0$, then the surface is concave upward. Therefore, the extremum is the minimum point of the surface. If we, next, introduce Equation 5.10 in Equation 5.3, we obtain the minimum error in the mean-square sense (see Problem 5.3.1):

$$J_{\min} = \sigma_d^2 - \mathbf{p}_{dx}^T \mathbf{w}^o = \sigma_d^2 - \mathbf{p}_{dx}^T \mathbf{R}_x^{-1} \mathbf{p}_{dx} \tag{5.11}$$

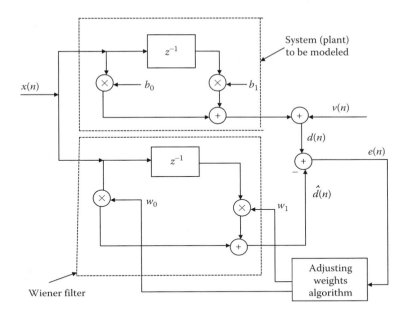

Figure 5.5

which indicates that the minimum point of the error surface is at a distance J_{min} above the **w**-plane. The above equation shows that if no correlation exists between the desired signal and the data, or equivalently $p_{dx} = 0$, the error is equal to the variance of the desired signal.

The problem we are facing is how to choose the length M of the filter. In the absence of a priori information, we compute the optimum coefficients, starting from a small reasonable number. As we increase the number, we check the minimum mean square error (MMSE) and if its value is small enough, e.g., MMSE < 0.01, we accept the corresponding number of the coefficients.

Example 5.2

We would like to find the optimum filter coefficients w_0 and w_1 of the Wiener filter which approximates (models) the unknown system with coefficients $b_0 = 1$ and $b_1 = 0.38$ (see Figure 5.5).

Solution

The following MATLAB program was used:

Book MATLAB m-file: ex5_3_1

```
%m-file: ex5_3_1
v=0.5*(rand(1,20)-0.5);%v=noise vector(20 uniformly
%distributed rv's with mean zero);
x=randn(1,20);% x=data vector entering
the system and
    %the Wiener
    %filter (20 normal distributed rv's with mean zero;
sysout=filter([1 0.38],1,x);% sysout=system output with
    %x as input;
    %filter(b,a,x) is a
    %MATLAB function, where b is the vector of the
    %coefficients of the ARMA numerator,
    % a is the vector of the coefficients of the
    %ARMA denominator;
dn=sysout+v;
rx1=xcorr(x,2,'biased');
rx=rx1(1,3:4);
Rx=toeplitz(rx);%toeplitz() is a MATLAB function that
                % gives the symmetric
                % autocorrelation matrix;
pdx=xcorr(x,dn,2, 'biased');%xcorr() a MATLAB function
                %that gives a symmetric biased
                %crosscorrelation;
p=pdx(1,2:3);
w=inv(Rx)*p';
dnsig1=xcorr(dn,1,'biased');
dnsig=dnsig1(2);
jmin=dnsig-p*w;                                        ■
```

Typical values found in this example are $\mathbf{R}_x$ = [0.9554 0.0295; 0.0295 0.9554]; $\mathbf{p}$ = [0.3413 0.9450]; $\mathbf{w}$ = [0.3270 0.9791]; J_{min} = 0.0179. We observe that the Wiener coefficients are close to the unknown system.

Orthogonality condition

In order for the set of filter coefficients to minimize the cost function $J(\mathbf{w})$, it is necessary and sufficient that the derivatives of $J(\mathbf{w})$ with respect to w_k be equal to zero for $k = 0, 1, 2, ..., M - 1$,

$$\frac{\partial J}{\partial w_k} = \frac{\partial}{\partial w_k} E\{e(n)e(n)\} = 2E\left\{e(n)\frac{\partial e(n)}{\partial w_k}\right\} = 0 \qquad (5.12)$$

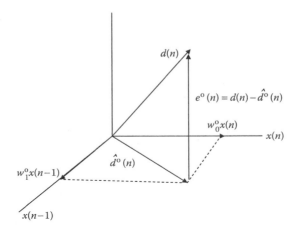

Figure 5.6

But

$$e(n) = d(n) - \sum_{m=0}^{M-1} w_m x(n-m) \tag{5.13}$$

and, hence, it follows (note that the derivative is for one variable w_k only and, therefore, the rest of the factors in the summation become zero besides one):

$$\frac{\partial e(n)}{\partial w_k} = -x(n-k) \tag{5.14}$$

Therefore, Equation 5.12 becomes

$$\boxed{E\{e^o(n)x(n-k)\} = 0, \qquad k = 0, 1, 2, ..., M-1} \tag{5.15}$$

where the superscript "o" denotes that the corresponding w_k's used to find the estimation error $e^o(n)$ are the optimal ones. Figure 5.6 illustrates the orthogonality principle where the error $e^o(n)$ is orthogonal (perpendicular) to the data set $\{x(n)\}$ when the estimator employs the optimum set of filter coefficients.

5.4   Wiener filtering examples

The following examples will elucidate the utility of the Wiener filtering.

Example 5.3 (filtering)

Filtering of noisy signals (noise reduction) is extremely important and the method has been used in many applications such as speech in noisy environment, reception of data across a noisy channel, enhancement of images, etc.

Let the received signal be $x(n) = d(n) + v(n)$, where $v(n)$ is noise with zero mean, variance σ_v^2 and it is uncorrelated with the desired signal, $d(n)$. Hence,

$$\begin{aligned}
p_{dx}(m) &= E\{d(n)x(n-m)\} = E\{d(n)d(n-m)\} \\
&\quad + E\{d(n)\}E\{v(n-m)\} = r_d(m) \\
&\quad + E\{d(n)\}0 = r_d(m)
\end{aligned} \tag{5.16}$$

Similarly, we obtain

$$r_x(m) = E\{x(n)x(n-m)\} = r_d(m) + r_v(m) \tag{5.17}$$

where we used the assumption that $d(n)$ and $v(n)$ are uncorrelated and $v(n)$ has zero mean value. Therefore, the Wiener–Hopf equation (Equation 5.9) becomes

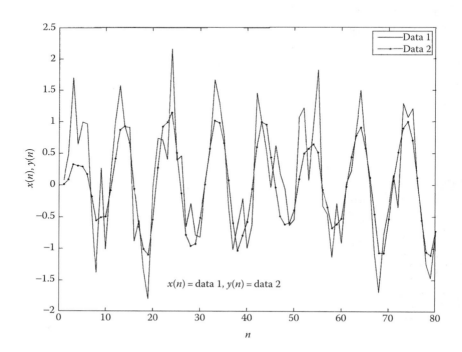

Figure 5.7

$$(\mathbf{R}_d + \mathbf{R}_v)\mathbf{w}^\circ = \mathbf{p}_{dx} \qquad (5.18)$$

The following Book MATLAB m-file was used to produce the results shown in Figure 5.7.

Book MATLAB m-file: ex5_4_1

```
%m—file:ex5_4_1
n=0:511;
d=sin(.2*pi*n);%desired signal
v=0.5*randn(1,512);%white Gaussian noise;
x=d+v;%input signal to Wiener filter;
rd1=xcorr(d,20,'biased');
rd=rd1(1,20+1:39);         %rdx=rd=biased autocorralation
                           %function of the
                           %desired signal;
rv1=xcorr(v,20,'biased');
rv=rv1(1,20+1:39);%rv=biased autoc. function of the
                           %noise;
R=toeplitz(rd(1,1:15))+toeplitz(rv(1,1:15));
pdx=rd(1,1:15);
w=inv(R)*pdx';
y=filter(w',1,x);%output of the filter;
```

But

$$\sigma_x^2 = \sigma_d^2 + \sigma_v^2, \quad \sigma_x^2 = r_x(0), \quad \sigma_d^2 = r_d(0), \quad \sigma_v^2 = r_v(0) \ (5.19)$$

and, hence, from the MATLAB function var(), we obtain var$(d) = $ var $(x) -$ var$(v) = 0.5332$ and $J_{\min} = 0.5332 - \mathbf{p}_{dx}\mathbf{w}^\circ = 0.0798$ ∎

Example 5.4 (system identification)

It is desired, using a Wiener filter, to estimate the unknown impulse response coefficients h_is of a FIR system (see Figure 5.8). The input $\{x(n)\}$ is a zero mean independent and identically distributed random variables (iid rvs) with variance σ_x^2. Let the impulse response $\mathbf{h}$ of the filter be $\mathbf{h} = [0.9 \ 0.6 \ 0.2]^T$. Since the input $\{x(n)\}$ is zero mean and iid rvs, the correlation matrix $\mathbf{R}_x$ is a diagonal matrix with elements having values σ_x^2. The desired signal $d(n)$ is the output of the unknown filter and it is given by (FIR system): $d(n) = 0.9x(n) + 0.6x(n-1) + 0.2x(n-2)$. Therefore, the cross-correlation output is given by

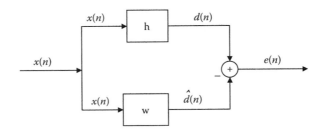

Figure 5.8

$$p_{dx}(i) = E\{d(n)x(n-i)\} = E\{[0.9x(n) + 0.6x(n-1)$$
$$+ 0.2x(n-2)]x(n-i)\}$$
$$= 0.9E\{x(n)x(n-i)\} + 0.6E\{x(n-1)x(n-i)\}$$
$$+ 0.2E\{x(n-2)x(n-i)\}$$
$$= 0.9r_x(i) + 0.6r_x(i-1) + 0.2r_x(i-2) \qquad (5.20)$$

Hence, we obtain $(r_x(m) = 0$, for $m \neq 0)$: $p_{dx}(0) = 0.9\sigma_x^2$, $p_{dx}(1) = 0.6\sigma_x^2$. The optimum filter is

$$\mathbf{w}^\circ = \mathbf{R}_x^{-1}\mathbf{p}_{dx} = (\sigma_x^2)^{-1}\begin{bmatrix} 1 & 0 \\ 0 & 1 \end{bmatrix}\begin{bmatrix} 0.9 \\ 0.6 \end{bmatrix} = (\sigma_x^2)^{-1}\begin{bmatrix} 0.9 \\ 0.6 \end{bmatrix}$$

and the MMSE is (assuming $\sigma_x^2 = 1$)

$$J_{min} = \sigma_d^2 - [0.9 \quad 0.6]\begin{bmatrix} 1 & 0 \\ 0 & 1 \end{bmatrix}\begin{bmatrix} 0.9 \\ 0.6 \end{bmatrix}.$$

But

$$\sigma_x^2 = E\{d(n)d(n)\} = E\{[0.9x(n) + 0.6x(n-1) + 0.2x(n-2)]^2\}$$
$$= 0.81 + 0.36 + 0.04 = 1.21, \text{ and, hence, } J_{min} = 1.21 - (0.9^2$$
$$+ 0.6^2) = 0.04.$$

Book MATLAB function for system identification using Wiener filter: [*w,jm*]=sspwieneridentif(*x, d, M*)

```
function [w, jm] =sspwieneridentif (x,d,M)
%function [w, jm] =sspwieneridentif (x,d,M) ;
%x=data entering both the unknown filter
% (system) and the Wiener filter;
%d=the desired signal=output of the unknown system;
%length (d) =length (x) ;
```

```
%M=number of coefficients of the Wiener filter;
%w=Wiener filter coefficients equal to M+1;
%jm=minimum mean square error;
pdx=xcorr(d,x,M,'biased');
p=pdx(1,M+1:2*M+1);
rx1=xcorr(x,M, 'biased');
rx=rx1(1,M+1:2*M+1);
R=toeplitz(rx);
w=inv(R)*p';
jm=var(d)-p*w;% var() is a MATLAB function;
```

By setting, for example, the following MATLAB procedure: x = randn (1,256); d = filter ([0.95 0.25
-0.4], 1, x); [w, jm] = sspwieneridentif(x, d, 4);,
we obtain **w** = [0.9497 0.2505 -0.3962 0.0031 0.0015], J_{min} =
0.0010. We note that, if we assume a larger number of
filter coefficients than those belonging to the unknown
system, the Wiener filter produces close approximate
values to those of the unknown system and, for the rest,
produces approximately zero values as it should.   ■

Example 5.5 (noise canceling)

In many practical applications there exists a need to cancel
the noise added to a signal. For example, we are talking in
a cell phone inside a car and the noise of the car, radio,
conversation by other riders, etc. are added to the message
we are trying to transmit. Similar circumstance appears
when pilots in planes and helicopter try to communicate,
or tank drivers try to do the same. Figure 5.9 shows
pictorially the noise contamination situations. Observe
that part of the noise is added to the signal to be trans-
mitted and another component is entering the Wiener

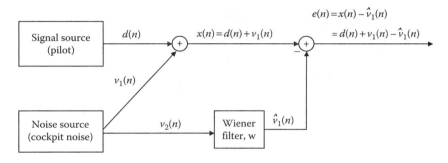

Figure 5.9

filter. Both components emanate from the same source but follow different paths in the same environment. This indicates that there is some degree of correlation between these two noises. It is assumed that the noises have zero mean values. The output of the Wiener filter will approximate the noise added to the desired signal and, thus, the error will be close to the desired signal. The Wiener filter in this case is

$$\mathbf{R}_{v_2}\mathbf{w}^o = \mathbf{p}_{v_1 v_2} \tag{5.21}$$

because the desired signal in this case is $\mathbf{v}_1$.

The individual components of the vector $\mathbf{P}_{v_1 v_2}$ are

$$
\begin{aligned}
p_{v_1 v_2}(m) &= E\{v_1(n)v_2(n-m)\} \\
&= E\{[x(n) - d(n)]v_2(n-m)\} \\
&= E\{x(n)v_2(n-m)\} - E\{d(n)v_2(n-m)\} \\
&= p_{xv_2}(m) \tag{5.22}
\end{aligned}
$$

Because $d(n)$ and $v_2(n)$ are uncorrelated,

$$E\{d(n)v_2(n-m)\} = E\{d(n)\}E\{v_2(n-m)\} = 0$$

since the noise has zero mean value. Therefore, Equation 5.21 becomes

$$\mathbf{R}_{v_2}\mathbf{w}^o = \mathbf{p}_{xv_2} \tag{5.23}$$

To demonstrate the effect of the Wiener filter, let $d(n) = 0.98^n \sin(0.2n\pi)$, $v_1(n) = 0.85v_1(n-1) + v(n)$, and $v_2(n) = -0.96v_2(n-1) + v(n)$, where $v(n)$ is the white noise with zero mean value and unit variance. The correlation matrix $\mathbf{R}_{v_2}$ and cross-correlation vector $\mathbf{p}_{xv_2}$ are found using the sample biased correlation function. Figure 5.10 shows the simulation results for three different-order filters using the Book MATLAB function given below.

Book MATLAB function for noise canceling:
[d,w,xn]=ssp_wiener_noisecancelor(dn,a1,a2,v,M,N)
```
function[d,w,xn]=ssp_wiener_noise_cancelor(dn,
  a1,a2,v,M,N)
%[d,w,xn]=ssp_wiener_noise_cancelor(dn,a1,a2,v,M,N);
%dn=desired signal;
```

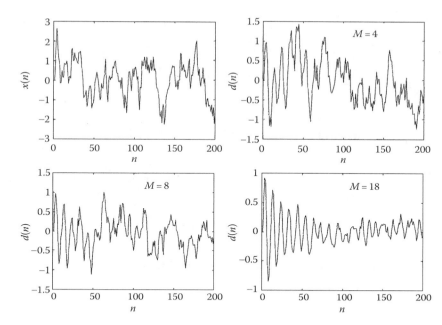

Figure 5.10

```
%a1=first order IIR coefficient,a2=first order
IIR   coefficient;
%v=noise;M=number of Wiener filter coefficients;N=number
%of sequence
%elemets of dn(desired signal) and v(noise);d=output
%desired signal;
%w=Wiener filter coefficients;xn=corrupted signal;
%en=xn—v1=d;
v1(1)=0;v2(1)=0;
for n=2:N
    v1(n)=a1*v1(n—1)+v(n—1);
    v2(n)=a2*v2(n—1)+v(n—1);
end;
v2autoc1=xcorr(v2,M,'biased');
v2autoc=v2autoc1(1,M+1:2*M+1);
xn=dn+v1;
Rv2=toeplitz(v2autoc);
p1=xcorr(xn,v2,'biased');
if M>N
    disp(['error:M must be less than N']);
end;
```

```
R=Rv2(1:M,1:M);
p=p1(1,(length(p1)+1)/2:(length(p1)+1)/2+M-1);
w=inv(R)*p';
yw=filter(w,1,v2);
d=xn-yw(:,1:N);
```
■

Self-correcting Wiener filter

It is proposed to arrange the standard single Wiener filter in a series form, each one having fewer coefficients by comparison to one-filter situation. This proposed configuration is shown in Figure 5.11. This configuration permits us to process the signal using filters with fewer coefficients, and permit us to stop at any desired stage, thus saving in computation.

> **Example 5.6 (linear prediction for an autoregressive (AR) process)**
>
> Plot the signal and the predicted one using Wiener filter.

Solution

> Consider the real-valued AR process of the second order
>
> $$x(n) = 0.92x(n-1) - 0.25x(n-2) + v(n) \qquad (5.24)$$
>
> where $v(n)$ is noise with variance 1 and zero mean value. With the help of MATLAB and for $N = 200$, we obtain

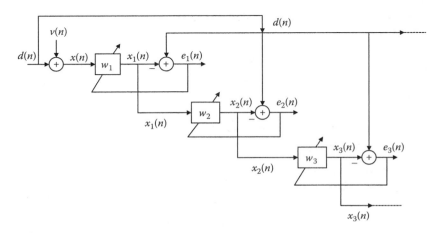

Figure 5.11

$$r_x(0) = 2.6626, \quad r_x(1) = 2.0612, \quad r_x(2) = 1.3283$$

Thus, the normal equations are

$$\begin{bmatrix} 2.6626 & 2.0612 \\ 2.0612 & 2.6626 \end{bmatrix} \begin{bmatrix} a_1 \\ a_2 \end{bmatrix} = - \begin{bmatrix} 2.0612 \\ 1.3283 \end{bmatrix}$$

or

$$\begin{bmatrix} a_1 \\ a_2 \end{bmatrix} = \begin{bmatrix} -1.3235 \\ 0.7097 \end{bmatrix}$$

Therefore, the predictor becomes

$$\hat{x}(n+1) = 1.3235x(n) - 0.7097x(n-1)$$

The signal $x(n)$ and the predictor $\hat{x}$ are shown in Figure 5.12 (in the figure $\hat{x} = xp$). In the legend, data 1 corresponds to $x(n)$ and data 2 corresponds to the predictor $xp(n)$. ■

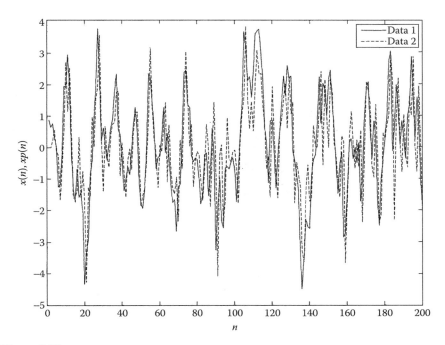

Figure 5.12

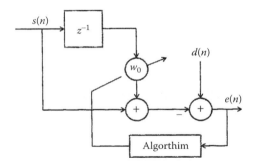

Figure 5.13

Problems

5.3.1 Verify Equation 5.11.

5.3.2 Find J_{min} using the orthogonality principle.

5.4.1 Find the cost function and the MSE surface for the two systems shown in Figure 5.13. Given: $E\{s^2(n)\} = 0.9$, $E\{s(n)s(n - 1)\} = 0.4$, $E\{d^2(n)\} = 3$, $E\{d(n)s(n)\} = -0.5$, and $E\{d(n)s(n - 1)\} = 0.9$.

Solutions, suggestions, and hints

5.3.1 $J(w) = J_{min} = \sigma_d^2 - 2w^{oT}R_x w^o = \sigma_d^2 - w^{oT}R_x w^o = \sigma_d^2 - (R_x^T w^o)^T w^o$

$$= \sigma_d^2 - (R_x w^o)^T w^o = \sigma_d^2 - p_{dx}^T w^o = \sigma_d^2 - p_{dx}^T R_x^{-1} p_{dx},$$

where R_x is symmetric and ws and ps are vectors.

5.3.2 $J = E\left\{e(n)\left[d(n) - \sum_{m=0}^{M-1} w_m x(n - m)\right]\right\} = E\{e(n)d(n)\} - \sum_{m=0}^{M-1} w_m E\{e(n)x(n - m)\}.$

If the coefficients have their optimum value, the orthogonality principle states that $E\{e(n)x(n - m)\} = 0$ and, hence,

$$J_{min} = E\{e(n)d(n)\} = E\left\{d(n)d(n) - \sum_{m=0}^{M-1} w_m^o x(n - m)d(n)\right\}$$

$$= \sigma_d^2 - \sum_{m=0}^{M-1} w_m^o E\{d(n)x(n - m)\} = \sigma_d^2 - \sum_{m=0}^{M-1} w_m^o p_{dx}(m)$$

$$= r_d(0) - p_{dx}^T w^o = r_d(0) - p_{dx}^T R_x^{-1} p_{dx}$$

5.4.1 $J = E\{(d(n) - [s(n) + w_0 s(n - 1)])^2\} = E\{d^2(n) - [s(n) + w_0 s(n - 1)]^2$

$$- d(n)[s(n) + w_0 s(n - 1)]\}$$

$$= E\{d^2(n)\} - E\{s^2(n)\} - w_0^2 E\{s^2(n - 1)\} - 2w_0 E\{s(n)s(n - 1)\}$$

$$- 2E\{d(n)s(n)\} - 2w_0 E\{s(n)s(n - 1)\}$$

$$= 3 - 0.9 - w_0^2 0.9 - 2w_0 0.4 + 2 \times 0.5 - 2w_0 0.4 = 3.1 - 0.9w_0^2$$

$$- 0.6w_0 \Rightarrow (\partial J/\partial w_0) = -0.9 \times 2w_0 - 1.6 = 0 \Rightarrow w_0 = 1.125$$

chapter six

Adaptive filtering—LMS algorithm

6.1 Introduction

In this chapter, we present the celebrated least mean square (LMS) algorithm, developed by Widrow and Hoff in 1960. This algorithm is a member of stochastic gradient algorithms, and because of its robustness and low computational complexity, it has been used in a wide spectrum of applications. This chapter introduces us the procedures of filtering random signals.

The LMS algorithm has the following most important properties:

1. It can be used to solve the Wiener–Hopf equation without finding matrix inversion. Furthermore, it does not require the availability of the autocorrelation matrix of the filter input and the cross-correlation between the filter input and its desired signal.
2. Its form is simple as well as its implementation, yet it is capable of delivering high performance during the adaptation process.
3. Its iterative procedure involves: (a) computing the output of an finite impulse response (FIR) filter produced by a set of tap inputs (filter coefficients), (b) generation of an estimated error by computing the output of the filter to a desired response, and (c) adjusting the tap weights (filter coefficients) based on the estimation error.
4. The correlation term needed to find the values of the coefficients at the $n + 1$ iteration contains the stochastic product $x(n)e(n)$ without the expectation operation that is present in the steepest descent method.
5. Since the expectation operation is not present, each coefficient goes through sharp variations (noise) during the iteration process. Therefore, instead of terminating at the Wiener solution, the LMS algorithm suffers random variation around the minimum point (optimum value) of the error-performance surface.
6. It includes a step-size parameter, μ, that must be selected properly to control stability and convergence speed of the algorithm.
7. It is stable and robust for a variety of signal conditions.

6.2 LMS algorithm

The LMS algorithm is given by

$$
\begin{aligned}
\mathbf{w}(n+1) &= \mathbf{w}(n) + 2\mu\mathbf{x}(n)[d(n) - \mathbf{x}^{\mathrm{T}}(n)\mathbf{w}(n)] \\
&= \mathbf{w}(n) + 2\mu\mathbf{x}(n)[d(n) - \mathbf{w}^{\mathrm{T}}(n)\mathbf{x}(n)] \\
&= \mathbf{w}(n) + 2\mu e(n)\mathbf{x}(n)
\end{aligned}
\tag{6.1}
$$

where

$$
y(n) = \mathbf{w}^{\mathrm{T}}(n)\mathbf{x}(n) \text{ is the filter output} \tag{6.2}
$$

$$
e(n) = d(n) - y(n) \text{ is the error} \tag{6.3}
$$

$$
\mathbf{w}(n) = [w_0(n) \quad w_1(n) \quad \dots \quad w_{M-1}(n)]^{\mathrm{T}} \text{ is the filter taps at time } n \tag{6.4}
$$

$$
\mathbf{x}(n) = [x(n) \quad x(n-1) \quad x(n-2) \quad \dots \quad x(n-M+1)]^{\mathrm{T}} \text{ is the input data} \tag{6.5}
$$

The algorithm defined by Equations 6.1 through 6.3 constitute the adaptive LMS algorithm. The algorithm at each iteration requires that $\mathbf{x}(n)$, $d(n)$, and $\mathbf{w}(n)$ are known. The LMS algorithm is a **stochastic gradient algorithm** if the input signal is a stochastic process. This results in varying the pointing direction of the coefficient vector during the iteration.

The key component of an adapter filter is the set of rules, or algorithms, that define how the term $2\mu e(n)\mathbf{x}(n)$ in Equation 6.1 is formed so that as $n \to \infty$ the filter coefficients converge to the Wiener–Hoff equation:

$$
\lim_{n \to \infty} \mathbf{w}(n) = \mathbf{R}_x^{-1}\mathbf{r}_{dx}
$$

Furthermore, it should not be necessary to know the second-order statistics of the input signal or the correlation between the input signal and the desired one. Finally, for nonstationary signals the filter should be able to adapt to the changing statistics and track the solution as it involves time.

An important consideration for the implementation of an adaptive filter is the requirement that the error signal, $e(n)$, be available to the adaptive algorithm. The error allows the filter to measure its performance and determine how the filter coefficients should be modified so that convergence occurs.

The simplicity of the LMS algorithm comes from the fact that the update of the kth coefficient,

$$
w_k(n+1) = w_k(n) + 2\mu e(n)x(n-k)
$$

requires only one multiplication and one addition (the value $2\mu e(n)$ needs only be computed once and may be used for all of the coefficients). Therefore, an LMS adaptive filter, having $M+1$ coefficients, requires $M+1$ multiplications and $M+1$ additions to update the filter coefficients. It avoids creating autocorrelation matrices and inversion of matrices.

A FIR adaptive filter realization is shown in Figure 6.1. Figure 6.2 shows the block diagram representation of the LMS filter and Table 6.1 presents the LMS algorithm.

Book MATLAB® LMS function: [w,y,e,J] = ssp_lms(x,dn,mu,M)

```
function[w,y,e,J]=ssp_lms(x,dn,mu,M)
%function[w,y,e,J]=aalms(x,dn,mu,M);
%all quantities are real-valued;
%x=input data to the filter; dn=desired signal;
%M=order of the filter;
%mu=step-size factor; x and dn must be of the same
%length;
N=length(x);
y=zeros(1,N); %initialized output of the filter;
w=zeros(1,M); %initialized filter coefficient vector;
for n=M:N
        x1=x(n:-1:n-M+1); %for each n the vector x1 is
                          %of length M with elements
                          %from x in reverse order;
    y(n)=w*x1';
    e(n)=dn(n)-y(n);
    w=w+2*mu*e(n)*x1;
end;
J=e.^2;%J is the learning curve of the adaptation;
```

The book MATLAB function ssp_lms_w.m, given below, provides the changing values of the filter coefficients as a function of the iteration number n:

Book MATLAB function providing the history of the filter coefficients: [w,y,e,J,w1] = ssp_lms_w(x,dn,mu,M)

```
function[w,y,e,J,w1]=ssp_lms_w(x,dn,mu,M)
%function[w,y,e,J,w1]=aalms1(x,dn,mu,M);
%this function provides also the changes of two
%filter coefficients
%versus iterations;
%all quantities are real-valued;
%x=input data to the filter; dn=desired signal;
```

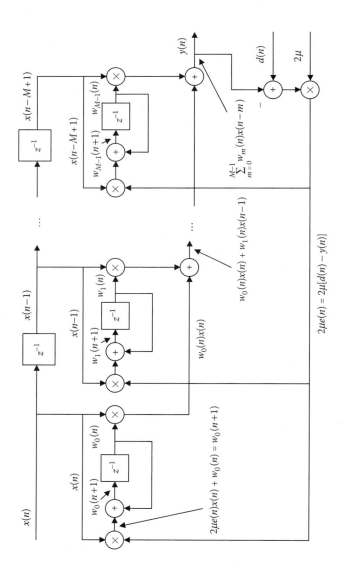

Figure 6.1

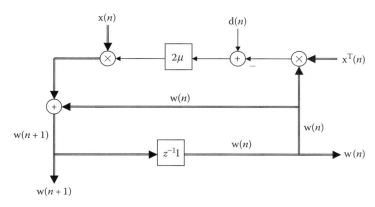

Figure 6.2

```
%M=order of the filter;
%mu=step size; x and dn must be of the same length;
%each column of the matrix w1 contains
%the history of each
%filter coefficient;
N=length(x);
y=zeros(1,N);
w=zeros(1,M); %initialized filter coefficient vector;
for n=M:N
    xl=x(n:-1:n-M+1); %for each n the vector xl of
                      %length M is produced
                      %with elements from x in
                      %reverse order;
```

Table 6.1 LMS Algorithm for a *M*th-Order
FIR Adaptive Filter

Inputs

M = filter length

μ = step-size factor

$\mathbf{x}(n)$ = input data to the adaptive filter

$\mathbf{w}(0)$ = initialization filter vector = $\mathbf{0}$

Outputs

$y(n)$ = adaptive filter output = $\mathbf{w}^{\mathrm{T}}(n)\mathbf{x}(n) \equiv \hat{d}(n)$

$e(n) = d(n) - y(n)$ = error

$\mathbf{w}(n + 1) = \mathbf{w}(n) + 2\mu e(n)\mathbf{x}(n)$

```
y(n)=w*x1';
e(n)=dn(n)-y(n);
w=w+2*mu*e(n)*x1;
w1(n-M+1,:)=w(1,:);
end;
J=e.^2;%J is the learning curve of the
%adaptive process; each column of the matrix w1
%depicts the history of each filter coefficient;
```

6.3   Examples using the LMS algorithm

The following examples will elucidate the use of the LMS algorithm to different areas of engineering and will create an appreciation for the versatility of this important algorithm.

Example 6.1 (linear prediction)

We can use an adaptive LMS filter as a predictor as shown in Figure 6.3a. Figure 6.3b shows the two-coefficient adaptive filter with its adaptive weight-control mechanism. The data $\{x(n)\}$ were created by passing a zero-mean white noise $\{v(n)\}$ through an autoregressive (AR) process described by the difference equation $x(n) = 0.4668x(n-1) - 0.3589x(n-2) + \text{randn}$. The LMS filter is used to predict the values of the AR filter parameters 0.4668 and −0.3589. A two-coefficient LMS filter predicts $x(n)$ by

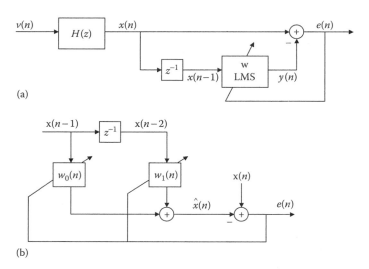

(a)

(b)

Figure 6.3

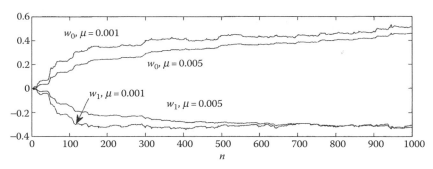

Figure 6.4

$$\hat{x}(n) = \sum_{i=0}^{1} w_i(n)x(n-1-i) \triangleq y(n) \qquad (6.6)$$

Figure 6.4 gives w_0 and w_1 versus the number of iterations for two different values of step size (μ = 0.001 and 0.0005). The adaptive filter is a two-coefficient filter. The noise is white and Gaussian distributed. The figure shows fluctuations in the values of coefficients as they converge to a neighborhood of their optimum value 0.4668 and −0.3589, respectively. As the step size μ becomes smaller, the fluctuations are not as large, but the convergence speed to the optimal values is slower.

Book MATLAB function for one-step LMS predictor:
[w,y,e,j,w1] = ssp_one_step_predictor(x,mu,M)

```
function[w,y,e,J,w1]=ssp_one_step_predictor(x,mu,M)
%function[w,y,e,J,w1]=ssp_one_step_predictor(x,mu,M);
%x=data=signal plus noise;mu=step size factor;
%M=number of filter
%coefficients;w1 is a matrix and each column
%is the history of each
%filter coefficient vesus time n;
N=length(x);
y=zeros(1,N);
w=zeros(1,M);
for n=M:N-1
    x1=x(n:-1:n-M+1);
    y(n)=w*x1';
```

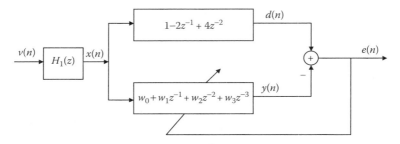

Figure 6.5

```
e(n)=x(n+1)-y(n);
w=w+2*mu*e(n)*x1;
w1(n-M+1,:)=w(1,:);
end;
J=e.^2;
%J is the learning curve of the adaptive process;
```

Example 6.2 (modeling)

Adaptive filtering can also be used to find the coefficients of an unknown system (filter). Let the unknown system be the one shown in Figure 6.5. The data $x(n)$ were produced in a similar way as in Example 6.1. The desired signal is given as follows: $d(n) = x(n) - 2x(n-1) + 4x(n-2)$. If the output is approximately equal to $d(n)$, it implies that the filter coefficients of the adaptive filter are approximately equal to the unknown system. Figure 6.6 shows the ability of adaptive filtering to identify the unknown system. After about 250 iterations, the system

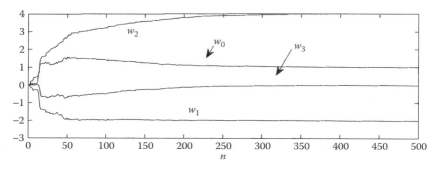

Figure 6.6

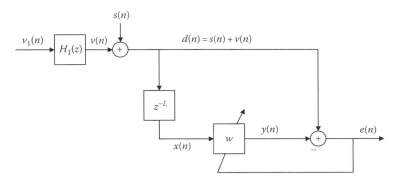

Figure 6.7

is practically identified. For this example, we used $\mu = 0.01$. It is apparent that the fourth coefficient is zero, as it should be, since the system to be identified has only three coefficients. ∎

Example 6.3 (noise cancellation)

A noise cancellation scheme is shown in Figure 6.7. We introduce in this example the following values: $H_1(z) = 1$ (or $h(n) = \delta(n)$), $v_1(n) =$ white noise $= v(n)$, $L = 1$, $s(n) = \sin(0.1\mu n)$. Therefore, the input signal to the filter is $x(n) = s(n-1) + v(n-1)$ and the desired signal is $d(n) = s(n) + v(n)$. The Book LMS MATLAB algorithm ssp_lms was used. Figure 6.8 shows the signal, the signal plus noise and the outputs of the adaptive filter for two different sets of coefficients: $M = 4$ and $M = 16$. The value of the μ was set equal to 0.001. ∎

Example 6.4 (inverse system identification)

To find the inverse of an unknown filter (system), we place the adaptive filter in series with the unknown system as shown in Figure 6.9. The delay is needed so that the system is causal. Figure 6.10 shows a typical learning curve. In this example, we used a three-coefficient IIR filter, and the input to the unknown system was a sine function with a white Gaussian noise. ∎

Example 6.5 (effect of μ on learning curve)

The following Book MATLAB function produces figures, as that shown in Figure 6.11:

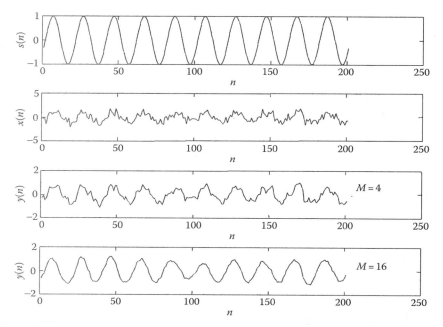

Figure 6.8

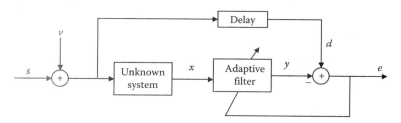

Figure 6.9

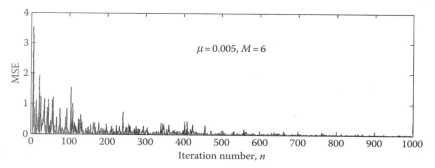

Figure 6.10

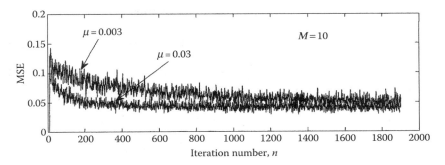

Figure 6.11

Book MATLAB function: [amse] = ssp_example635(mu,M,an)

```
function[amse]=ssp_example639(mu,M,an)
%function[amse]=ssp_example639(mu,M);
%M=number of filter coefficients;an=number of times
%the error to be averaged;
eav=zeros(1,2000);
dn(1)=0;dn(2)=0;x(1)=0;x(2)=0;
for k=1:an
  for n=3:2000
    dn(n)=0.9621*dn(n-1)-0.2113*dn(n-2)+0.2*randn;
    x(n)=dn(n-1);
  end;
  [w,y,e]=ssp_lms(x,dn,mu,M);
  eav=eav+e.^2;
end;
amse=eav/an;
```

6.4   *Properties of the LMS method*

1. The steepest-descent method reaches the minimum mean-square error J_{min} as $n \to \infty$ and $\mathbf{w}(n) \to \mathbf{w}^{\circ}$ which is identical to Wiener solution.
2. The LMS method produces an error $J(\infty)$ that approaches J_{min} as $n \to \infty$ and remains larger than J_{min}.
3. The LMS method produces a $\mathbf{w}(n)$, as the iterations $n \to \infty$, that is close to the optimum $\mathbf{w}^{\circ}$.
4. The steepest-descent method has a well-defined learning curve consisting of a sum of decaying exponentials.
5. The LMS learning curve is a sum of noisy decaying exponentials and the noise, in general, decreases the smaller values the step-size parameter μ takes.

6. In the steepest-descent method, the correlation matrix $\mathbf{R}_x$ of the data $\mathbf{x}(n)$ and the cross-correlation vector $\mathbf{p}_{dx}(n)$ are found using ensemble averaging operations from the realizations of the data $\mathbf{x}(n)$ and desired signal $d(n)$.

7. In the LMS filter, an ensemble of learning curves is found under identical filter parameters and then averaged point by point.

It is recommended that the readers repeat the examples by varying all the different parameters, the noise amplitude, the signal amplitude, etc. so that it requires experience of the sensitivity of these factors to obtain a solution.

chapter seven

Adaptive filtering with variations of LMS algorithm

7.1 Sign algorithms

This chapter covers the most popular modified least mean square (LMS)-type algorithms proposed by researchers over the past years as well as some recent ones proposed by Poularikas and Ramadan.* Most of these algorithms were designed on an ad hoc basis to improve convergence behavior, reduce computational requirements, decrease the steady-state mean-square error (MSE), etc. We start this chapter by introducing, first, the sign algorithms.

7.1.1 Error sign algorithm

The error sign algorithm is defined by

$$\mathbf{w}(n+1) = \mathbf{w}(n) + 2\mu \, \text{sign}(e(n))\mathbf{x}(n) \tag{7.1}$$

where

$$\text{sign}(n) = \begin{cases} 1, & n > 0 \\ 0, & n = 0 \\ -1, & n = -1 \end{cases} \tag{7.2}$$

is the signum function. By introducing the signum function, and setting μ to a value of power of 2, the hardware implementation is highly simplified (shift and add/subtract operation only).

Book MATLAB® function for the sign algorithm:
[w,y,e,J,w1] = ssp_lms_sign(x,dn,mu,M)

```
function [w,y,e,J,w1]=ssp_lms_sign(x,dn,mu,M)
%function [w,y,e,J,w1]=ssp_lms_sign(x,dn,mu,M);
%all quantities are real-valued;
%x=input data to the filter;dn=desired signal;
%M=order of the filter;
```

* A. D. Poularikas and Z. Ramadan, *Adaptive Filtering Primer*, Taylor & Francis, Boca Raton, FL, 2006.

```
%mu=step size parameter;x and dn must be of the same length
N=length(x);
y=zeros(1,N);
w=zeros(1,M);%initialized filter coefficient vector
for n=M:N
    x1=x(n:-1:n-M+1);%for each n the vector x1 is produced
    %of length M with elements from x in reverse order;
    y(n)=w*x1';
    e(n)=dn(n)-y(n);
    w=w+2*mu*sign(e(n))*x1;
    w1(n-M+1,:)=w(1,:);
end;
J=e.^2;
%the columns of w1 depict the history of the filter
%coefficients;
```

Figure 7.1 shows the noise reduction using the sign algorithm. For this case we set the following values: $M = 30$, $\mu = 0.001$, the input signal to the filter $x(n) = 0.99^n \sin(0.1n\pi) + 0.5$ randn, the desired signal $d(n) = x(n-1)$, and the output signal $y(n)$.

7.1.2   Normalized LMS (NLMS) sign algorithm

The NLMS sign algorithm is

$$\mathbf{w}(n+1) = \mathbf{w}(n) + 2\mu \frac{\text{sign}(e(n))\mathbf{x}(n)}{\varepsilon + \|\mathbf{x}(n)\|^2} \tag{7.3}$$

Book MATLAB function for normalized LMS sign algorithm

```
function[w,y,e,J,w1]=ssp_normalized_lms_sign(x,dn,mu,M,ep)
%function[w,y,e,J,w1]=ssp_normalized_lms_sign(x,dn,mu,M,ep);
%all quantities are real-valued;
```

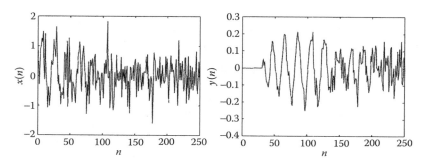

Figure 7.1

```
%x=input data to the filter;dn=desired signal;
%M=order of the filter;
%mu=step size parameter;x and dn must be of the same
%length;
%ep=small;
N=length(x);
y=zeros(1,N);
w=zeros(1,M);%initialized filter coefficient vector
for n=M:N
    x1=x(n:-1:n-M+1);%for each n the vector x1 is produced
    %of length M with elements from x in reverse order;
    y(n)=w*x1';
    e(n)=dn(n)-y(n);
    w=w+2*mu*sign(e(n))*x1/(ep+x1*x1');
    w1(n-M+1,:)=w(1,:);
end;
J=e.^2;
%the columns of w1 depict the history of the filter
%coefficients;
```

Figure 7.2 shows the noise removal from a decaying sine signal. The following factors and constants were used:

$$s = 0.99^n \sin(0.1n\pi), \quad x = s + 0.5\,\text{randn}, \quad d(n) = x(n-1),$$
$$\mu = 0.001, \quad M = 30, \quad ep = 0.1$$

7.1.3   Signed-regressor algorithm

The signed-regressor or data-sign algorithm is given as follows:

$$\mathbf{w}(n+1) = \mathbf{w}(n) + 2\mu e(n)\,\text{sign}(\mathbf{x}(n)) \tag{7.4}$$

where the sign function is applied to $x(n)$ on element-by-element basis.

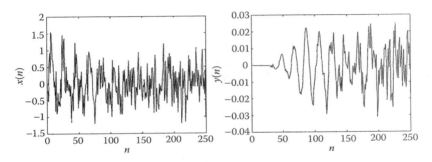

Figure 7.2

7.1.4   *Error-normalized signed-regressor algorithm*

The error-normalized signed-regressor algorithm is

$$\mathbf{w}(n+1) = \mathbf{w}(n) + 2\mu \frac{e(n)\mathrm{sign}(\mathbf{x}(n))}{\varepsilon + \|\mathbf{en}\|^2} \qquad (7.5)$$

where $\mathbf{en} = [e(n) \quad e(n-1) \quad e(n-2) \quad \ldots]^\mathrm{T}$.

Book MATLAB function for error normalized signed-regressor algorithm:
[w,y,e,J,w1] = ssp_lms_norm_signed_regressor(x,dn,mu,M,ep)

```
function [w,y,e,J,w1]=ssp_lms_norm_signed_ ...
   regressor(x,dn,mu,M,ep)
%function [w,y,e,J,w1]=ssp_lms_norm_signed_
%regressor(x,dn,mu,M);
%all quantities are real-valued;
%x=input data to the filter;dn=desired signal;
%M=order of the filter;
%mu=step size;x and dn must be of the same length
N=length(x);
y=zeros(1,N);
w=zeros(1,M);%initialized filter coefficient vector
en=0;
for n=M:N
    x1=x(n:-1:n-M+1);%for each n the vector x1 is
    %of length M with elements from x in reverse order;
    y(n)=w*x1';
    e(n)=dn(n)-y(n);
    en=en+(abs(e(n)))^2;
    w=w+2*mu*e(n)*sign(x1)/(ep+en); %ep=a small positive
    %number e.g. 0.05;
  w1(n-M+1,:)=w(1,:);
end;
J=e.^2;
%the columns of w1 depict the history of the filter
%coefficients;
```

Figure 7.3 shows the noise reduction using the above algorithm. The different factors and constants are

$$s(n) = 2\sin(0.1n\pi)0.99^n, \quad x = s + \mathrm{randn}, \quad \mu = 0.005,$$
$$M = 18, \quad \mathrm{ep} = 0.1, \quad d(n) = x(n-1)$$

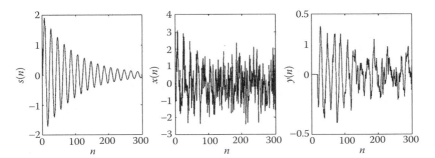

Figure 7.3

7.1.5   Sign–sign algorithm

The sign–sign algorithm is given by

$$\mathbf{w}(n+1) = \mathbf{w}(n) + 2\mu\,\text{sign}(e(n))\text{sign}(\mathbf{x}(n)) \qquad (7.6)$$

Book MATLAB function for sign-sign algorithm

```
function[w,y,e,J,w1]=ssp_lms_sign_sign(x,dn,mu,M)
%function[w,y,e,J,w1]=ssp_lms_sign_sign(x,dn,mu,M)
%all quantities are real-valued;
%x=input data to the filter;dn=desired signal;
%M=order of the filter;
%mu=step size parameter;x and dn must be of the same
%length
N=length(x);
y=zeros(1,N);
w=zeros(1,M);%initialized filter coefficient vector
for n=M:N
    x1=x(n:-1:n-M+1);%for each n the vector x1
    %is produced
    %of length M with elements from x in reverse order;
    y(n)=w*x1';
    e(n)=dn(n)-y(n);
    w=w+2*mu*sign(e(n))*sign(x1);
    w1(n-M+1,:)=w(1,:);
end;
J=e.^2;
%the columns of w1 depict the history of the filter
%coefficients;
```

7.1.6   Error-data-normalized sign–sign algorithm

The error-data-normalized algorithm is

$$\mathbf{w}(n+1) = \mathbf{w}(n) + 2\mu \frac{\text{sign}(e(n))\text{sign}(\mathbf{x}(n))}{\alpha \parallel \mathbf{en} \parallel^2 + \gamma \parallel \mathbf{x}(n) \parallel^2} \tag{7.7}$$

where $\mathbf{en} = [e(n) \quad e(n-1) \quad e(n-2) \quad \dots]^{\mathrm{T}}$.

Book MATLAB function for the error-data normalized sign-sign algorithm:
[w,y,e,J,w1] = ssp_lms_error_data_sign_sign(x,dn,mu,al,ga)

```
function [w,y,e,J,w1]=ssp_lms_error_data_sign_
  sign(x,dn,mu,M,al,ga)
%function [w,y,e,j,w1]=ssp_lms_error_data_sign_
%sign(x,dn,mu,M,al,ga);
%all quantities are real; x=input data to the filter;
%dn=desired signal;M=filter order;mu=step size;x and dn
%must have the
%same length;0<al<1; ga=1-al, or diferent value less
%than one;
N=length(x);
y=zeros(1,N);
w=zeros(1,M);%initialization;
en=0;
for n=M:N
    x1=x(n:-1:n-M+1);
    y(n)=w*x1';
    e(n)=dn(n)-y(n);
    en=en+(abs(e(n)))^2;
    w=w+2*mu*en*(sign(e(n))*sign(x1))/(al*en+ga*x1*x1');
    w1(n-M+1,:)=w(1,:);
end;
J=e.^2;
%columns of w1 depict the history of the filter
%coefficients;
```

7.2   NLMS algorithm

Consider the LMS recursion algorithm

$$\mathbf{w}(n+1) = \mathbf{w}(n) + 2\mu(n)e(n)\mathbf{x}(n) \tag{7.8}$$

where the step-size parameter $\mu(n)$ varies in time. It has been shown in the literature that the stability, convergence, and steady-state behavior of the LMS algorithm are influenced by the filter length and the power of the signal. Therefore, we can set

$$\mu(n) = \frac{1}{2\mathbf{x}^T(n)\mathbf{x}(n)} = \frac{1}{2||\mathbf{x}(n)||^2} \tag{7.9}$$

in Equation 7.8 to find the recursion

$$\mathbf{w}(n+1) = \mathbf{w}(n) + \frac{1}{\mathbf{x}^T(n)\mathbf{x}(n)} e(n)\mathbf{x}(n) \tag{7.10}$$

It is recommended in practice to use a more relaxed recursion that guarantees reliable results. Hence, we write

$$\mathbf{w}(n+1) = \mathbf{w}(n) + \frac{\bar{\mu}}{\varepsilon + \mathbf{x}^T(n)\mathbf{x}(n)} e(n)\mathbf{x}(n) \tag{7.11}$$

where $\bar{\mu}$ and ε are constants. The small constant ε prevents division by a very small number of the data norm. The NLMS algorithm is shown in Table 7.1.

Table 7.1   NLMS Algorithm

Real-Valued Functions	Complex-Valued Functions
Input Initialization vector: $\mathbf{w}(n) = 0$ Input vector: $\mathbf{x}(n)$ Desired output: $d(n)$ Step-size parameter: $\bar{\mu}$ Constant: ε Filter length: M	
Output Filter output: $y(n)$ Coefficient vector: $\mathbf{w}(n+1)$	
Procedure 1. $y(n) = \mathbf{w}^T(n)\mathbf{x} = \mathbf{w}(n)\mathbf{x}^T(n)$ 2. $e(n) = d(n) - y(n)$ 3. $\mathbf{w}(n+1) = \mathbf{w}(n) + \dfrac{\bar{\mu}}{\varepsilon + \mathbf{x}^T(n)\mathbf{x}(n)} e(n)\mathbf{x}(n)$	1. $y(n) = \mathbf{w}^H(n)\mathbf{x}(n)$ 2. $e(n) = d(n) - \mathbf{w}^H(n)\mathbf{x}(n)$ 3. $w(n+1) = w(n) + \dfrac{\bar{\mu}}{\varepsilon + \mathbf{x}^H(n)\mathbf{x}(n)} \mathbf{x}(n)e^*(n)$

Note: The superscript H stands for Hermitian, or equivalently conjugate transpose.

Book MATLAB function for normalized LMS and complex data:
[w,y,e,J,w1] = ssp_normalized_complex_lms(x,dn,mubar,M,c)

```
function[w,y,ek,J,w1]=ssp_normalized_complex_
  lms(x,dn,mubar,M,c)
%function[w,y,e,J,w1]=ssp_normalized_complex_
%lms(x,dn,mubar,M,c)
%x=input data to the filter;dn=desired signal;
%M=filter order;c=constant;mubar=step-size equivalent
%parameter;
%x and dn must be of the same length;J=learning curve;
N=length(x);
y=zeros(1,N);
w=zeros(1,M);%initialized filter coefficient vector;
for n=M:N
    x1=x(n:-1:n-M+1);%for each n vector x1 is of length
    %M with elements from x in reverse order;
    y(n)=conj(w)*x1';
    e(n)=dn(n)-y(n);
    w=w+(mubar/(c+conj(x1)*x1'))*conj(e(n))*x1;
    w1(n-M+1,:)=w(1,:);
end;
J=e.^2;
%the columns of the matrix w1 depict the history of the
%filter coefficients;
```

Figure 7.4 shows the results of the NLMS filter with the following constants and factors:

$$s = 2\sin(0.1n\pi)0.99^n, \quad x = s + 0.5\,\text{randn}, \quad \bar{\mu} = 0.5, \quad \varepsilon = 0.1, \quad d(n) = x(n-1)$$

Appendix C develops the NLMS algorithm.

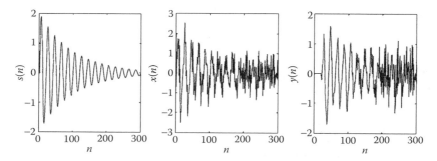

Figure 7.4

7.3   Variable step-size LMS (VSLMS) algorithm

The VSLMS algorithm was introduced to facilitate the conflicting require-ments, whereas a large step-size parameter is needed for fast convergence and a small step-size parameter is needed to reduce the misadjustment factor (the factor is proportional to the difference between the MSE as $n \to \infty$ and the minimum square error based on Wiener's solution). When the adaptation begins and $\mathbf{w}(n)$ is far from its optimum value, the step-size parameter should be large in order for the convergence to be rapid. However, as the filter value $\mathbf{w}(n)$ approaches the steady-state solution, the step-size parameter should decrease in order to reduce the excess MSE. To accomplish the variation of the step-size parameter, each filter coefficient is given a separate time-varying step-size parameter such that the LMS recursion algorithm takes the form:

$$w_i(n+1) = w_i(n) + 2\mu_i(n)e(n)x(n-i), \quad i = 0, 1, \ldots, M-1 \qquad (7.12)$$

Step sizes are determined in an ad hoc manner, based on monitoring sign changes in the instantaneous gradient estimate $e(n)x(n-i)$. It was argued that successive changes in the sign of the gradient estimate indicate that the algorithm is close to its optimal solution and, hence, the step-size value must be decreased. The reverse is also true. The decision of decreasing the value of the step-size by some factor c_1 is based on some number m_1 suc-cessive changes of $e(n)x(n-i)$. Increasing the step-size parameter by some factor c_2 is based on m_2 successive sign changes. The parameters m_1 and m_2 can be adjusted to optimize performance, as can the factors c_1 and c_2.

   The set of update equations (Equation 7.12) may be written in the matrix form:

$$\mathbf{w}(n+1) = \mathbf{w}(n) + 2\mu(n)e(n)\mathbf{x}(n) \qquad (7.13)$$

where $\mu(n)$ is a diagonal matrix with the following elements in its diagonal: $\mu_0(n)$, $\mu_1(n)$, $\ldots$, $\mu_{M-1}(n)$. The VSLMS algorithm is given in Table 7.2.

7.4   Leaky LMS algorithm

It turns out that when one of the eigenvalues of the correlation matrix is zero, the solution may become unstable. Therefore, it is important that we proceed to stabilize the LMS algorithm. One way to remedy this difficulty is to intro-duce a **leakage coefficient** γ into the LMS algorithm as follows:

$$\mathbf{w}(n+1) = (1-2\mu\gamma)\mathbf{w}(n) + 2\mu e(n)\mathbf{x}(n) \qquad (7.14)$$

where $0 < y \ll 1$. The leakage coefficient introduces a constraint into step-size parameter and, thus, guarantees convergence.

Table 7.2   VSLMS Algorithm

Input

Initial coefficient vector: $\mathbf{w}(0)$

Input data vector: $\mathbf{x}(n) = [x(n) \quad x(n-1) \quad \ldots \quad x(n-M+1)]^T$

Gradient term: $g_0(n-1) = e(n-1)x(n-1)$, $g_1(n-1) = e(n-1)x(n-1)$, ...,
$$g_{M-1}(n-1) = e(n-1)x(n-M)$$

Step-size parameter: $\mu_0(n-1)$, $\mu_1(n-1)$, ..., $\mu_{M-1}(n-1)$
$$a = \text{small positive constant}$$
$$\mu_{\max} = \text{positive constant}$$

Outputs

Desired output: $d(n)$

Filter output: $y(n)$

Filter update: $\mathbf{w}(n+1)$

Gradient term: $g_0(n), g_1(n), ..., g_{M-1}(n)$

Update step-size parameter: $\mu_0(n), \mu_1(n), ..., \mu_{M-1}(n)$

Execution

  1. $y(n) = \mathbf{w}^T(n)\mathbf{x}(n)$
  2. $e(n) = d(n) - y(n)$
  3. Weights and step-size parameter adaptation:
     For $i = 0, 1, 2, ..., M-1$
     $g_i(n) = e(n)x(n-i)$
     $\mu_i(n) = \mu_i(n-1) + \text{assign}(g_i(n))\text{sign}(g_i(n))$
     if $\mu_i(n) > \mu_{\max}$, $\mu_i(n) = \mu_{\max}$
     if $\mu_i(n) < \mu_{\min}$, $\mu_i = \mu_{\min}$
     $w_i(n+1) = w_i(n) + 2\mu_i(n)g_i(n)$
     end

Book MATLAB leaky MATLAB function:
[w,y,e,J,w1] = ssp_leaky_lms(x,dn,mu,gamma,M)

```
function[w,y,e,J,w1]=ssp_leaky_lms(x,dn,mu,gama,M)
%function[w,y,e,J,w1]=ssp_leaky_lms(x,dn,mu,gama,M)
%all signals are real valued;x=input to filter;
%y=output from the filter;dn=desired signal;
%mu=step-size factor;gama=gamma factor<<1;
%M=number of filter coefficients;w1=matrix whose M
%rows give the history of each filter coefficient;
N=length(x);
y=zeros(1,N);
w=zeros(1,M);
for n=M:N
```

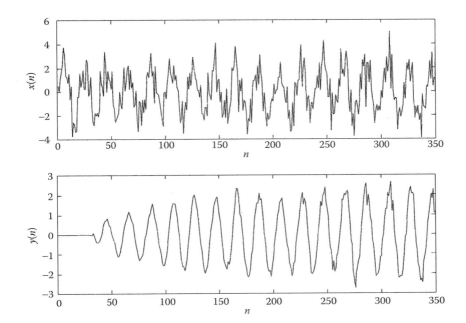

Figure 7.5

```
    x1=x(n:-1:n-M+1);
    y(n)=w*x1';
    e(n)=dn(n)-y(n);
    w=(1-2*mu*gama)*w+2*mu*e(n)*x1;
    w1(n-M+1,:)=w(1,:);
end;
J=e.^2;
```

$$J_c = E\{e^2(n)\} + \lambda(\mathbf{c}^T\mathbf{w} - a) \tag{7.15}$$

Figure 7.5 shows the result of the leaky algorithm with the following factors and constants:

$$s = 2\sin(0.1n\pi), \quad x = s + \text{randn}, \quad d(n) = x(n-1), \quad \mu = 0.0005, \quad \gamma = 0.1, \quad M = 30$$

7.5 Linearly constrained LMS algorithm

In all the analyses done for Wiener filtering problem—steepest-descent method, Newton's method, and LMS algorithm—no constrain was imposed on the solution of minimizing the MSE. However, in some applications there might be some mandatory constraints that must be taken into consideration in solving optimization problems. In this section, we discuss the filtering problem of minimizing the MSE subject to a general constraint.

The error between the desired signal and the output of the filter is

$$e(n) = d(n) - \mathbf{w}^T(n)\mathbf{x}(n) \tag{7.16}$$

We wish to minimize this error in the mean-square sense, subject to the constraint

$$\mathbf{c}^T\mathbf{w} = a \tag{7.17}$$

where a is a constant and $\mathbf{c}$ is a fixed vector. Using the Lagrange multiplier method, we write

$$J_c = E\{e^2(n)\} + \lambda(\mathbf{c}^T\mathbf{w} - a) \tag{7.18}$$

where λ is the Lagrange multiplier. Hence, the following relations

$$\nabla_w J_c = 0 \quad \text{and} \quad \frac{\partial J_c}{\partial \lambda} = 0 \tag{7.19}$$

must be satisfied simultaneously. The term $\partial J_c/\partial \lambda$ produces the constraint (Equation 7.17). Next, we substitute the error $e(n)$ in Equation 7.18 to obtain

$$J_c = J_{min} + \mathbf{z}^T\mathbf{R}_x\mathbf{z} + \lambda(\mathbf{c}^T\mathbf{z} - a') \tag{7.20}$$

where

$$\mathbf{z}(n) = \mathbf{w}(n) - \mathbf{w}^\circ, \quad \mathbf{w}^\circ = \mathbf{R}_x^{-1}\mathbf{p}_{dx}, \quad \mathbf{R}_x = E\{\mathbf{x}(n)\mathbf{x}^T(n)\} \tag{7.21}$$

and

$$\mathbf{p}_{dx} = E\{d(n)\mathbf{x}(n)\}, \quad a' = a - \mathbf{c}^T\mathbf{W}^\circ \tag{7.22}$$

The solution now has changed to $\nabla_z J = 0$ and $\partial J_c/\partial \lambda = 0$. Hence, from Equation 7.20 we obtain

$$\nabla_z J_c = \begin{bmatrix} \dfrac{\partial J_c}{\partial \mathbf{z}_1} \\ \vdots \\ \dfrac{\partial J_c}{\partial \mathbf{z}_M} \end{bmatrix} = \begin{bmatrix} 2\mathbf{z}_1 r_1 + 2\mathbf{z}_2 r_2 + \cdots + 2\mathbf{z}_M r_M \\ \vdots \\ 2\mathbf{z}_1 r_M + 2\mathbf{z}_2 r_{M-1} + \cdots 2\mathbf{z}_M r_1 \end{bmatrix} + \lambda \begin{bmatrix} c_1 \\ \vdots \\ c_M \end{bmatrix} = \begin{bmatrix} 0 \\ \vdots \\ 0 \end{bmatrix} \triangleq 0 \tag{7.23}$$

or in matrix form:

$$2\mathbf{R}_x\mathbf{z}_c^o + \lambda\mathbf{c} = 0 \tag{7.24}$$

where $\mathbf{z}_c^o$ is the constraint optimum value of the vector $\mathbf{z}$. In addition, the constraint gives the relation:

$$\frac{\partial J_c}{\partial\lambda} = \mathbf{c}^T\mathbf{z}_c^o = 0 \tag{7.25}$$

Solve the system of the last equations for λ and $\mathbf{z}_c^o$ we obtain

$$\lambda = -\frac{2a'}{\mathbf{c}^T\mathbf{R}_x^{-1}\mathbf{c}}, \quad \mathbf{z}_c^o = \frac{a'\mathbf{R}_x^{-1}\mathbf{c}}{\mathbf{c}^T\mathbf{R}_x^{-1}\mathbf{c}} \tag{7.26}$$

Substituting the value of λ in Equation 7.20, we obtain the minimum value of J_c to be

$$J_c = J_{min} + \frac{a'^2}{\mathbf{c}^T\mathbf{R}_x^{-1}\mathbf{c}} \tag{7.27}$$

But $\mathbf{w}(n) = \mathbf{z}(n) + \mathbf{w}^o$ and, hence, using Equation 7.26 we obtain

$$\mathbf{w}_c^o = \mathbf{w}^o + \frac{a'\mathbf{R}_x^{-1}\mathbf{c}}{\mathbf{c}^T\mathbf{R}_x^{-1}\mathbf{c}}$$

Note: The second term of Equation 7.27 is the excess MSE produced by the constraint.

To obtain the recursion relation subject to constraint 7.17, we must proceed in two steps:

Step 1:

$$\mathbf{w}'(n) = \mathbf{w}(n) + 2\mu e(n)\mathbf{x}(n) \tag{7.28}$$

Step 2:

$$\mathbf{w}(n+1) = \mathbf{w}'(n) + \mathbf{h}(n) \tag{7.29}$$

where $\mathbf{h}(n)$ is chosen so that $\mathbf{c}^T\mathbf{w}(n+1) = a$ while $\mathbf{h}^T(n)\mathbf{h}(n)$ is minimized. In other words, we choose the vector $\mathbf{h}(n)$ so that Equation 7.17 holds after step 2, while the perturbation introduced by $\mathbf{h}(n)$ is minimized. The problem can be solved using Lagrange multiplier method that gives

$$\mathbf{h}(n) = \frac{a - \mathbf{c}^T\mathbf{w}'(n)}{\mathbf{c}^T\mathbf{c}}\mathbf{c} \tag{7.30}$$

Table 7.3   LMS Algorithm Linearly
Constrained

Input
Initial coefficient vector: $\mathbf{w}(0) = \mathbf{0}$
Input data vector: $\mathbf{x}(n)$
Desired output: $d(n)$
Constant vector: $\mathbf{c}$
Constraint constant: a

Output
Filter output: $y(n)$

Procedure
$y(n) = \mathbf{w}^{\mathrm{T}}(n)\mathbf{x}(n)$
$e(n) = d(n) - y(n)$
$\mathbf{w}'(n) = \mathbf{w}(n) + 2\mu e(n)\mathbf{x}(n)$
$\mathbf{w}(n + 1) = \mathbf{w}'(n) + \dfrac{a - \mathbf{c}^{\mathrm{T}}\mathbf{w}'(n)}{\mathbf{c}^{\mathrm{T}}\mathbf{c}}\mathbf{c}$

Thus, we obtain the final form of Equation 7.29 to be

$$\mathbf{w}(n + 1) = \mathbf{w}'(n) + \frac{a - \mathbf{c}^{\mathrm{T}}\mathbf{w}'(n)}{\mathbf{c}^{\mathrm{T}}\mathbf{c}} \qquad (7.31)$$

The constraint algorithm is given in Table 7.3.

Book MATLAB function for constrained LMS algorithm:
[w,e,y,J,w2] = ssp_constrained_lms(x,dn,c,a,mu,M)

```
function[w,e,y,J,w2]=ssp_constrained_lms(x,dn,c,a,mu,M)
%function[w,e,y,J,w2]=ssp_constrained_
%lms(x,dn,c,a,mu,M);
%x=data vector;dn=desired vector of equal length
%with x;
%c=constant row vector of length M;a=constant, e.g.
%a=0.8;mu=step-
%size parameter;M=filter order(number of filter
%coefficients);
%w2=matrix whose columns give the history of each
%coefficient;
w=zeros(1,M);
N=length(x);
```

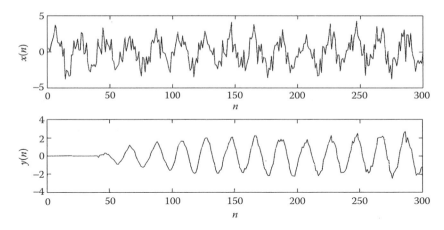

Figure 7.6

```
for n=M:N;
    y(n)=w*x(n:-1:n-M+1)';
    e(n)=dn(n)-y(n);
    w1=w+2*mu*e(n)*x(n:-1:n-M+1);
    w=w1+((a-c*w1')*c/(c*c'));
    w2(n-M+1,:)=w(1,:);
end;
J=e.^2;
```

Figure 7.6 shows the result of a constrained LMS algorithm. The factors and constants are

$$s(n) = 2\sin(0.1n\pi), \quad x(n) = s(n) + \text{randn}, \quad c = \text{ones}(1, M), \quad a = 0.6,$$
$$\mu = 0.0005, \quad M = 40, \quad d(n) = x(n-1)$$

7.6   Self-correcting adaptive filtering (SCAF)

One way by which we may improve the output of the adaptive filter, so that it is approximately equal to the desired one, is to use the proposed **self-correcting adaptive filtering** as shown in Figure 7.7. In this proposed configuration, the desired signal is compared with signals which become closer and closer to the desired one. The output of the *i*th stage is related to the previous one as follows:

$$y_{i+1}(n) = y_i^* w_{i+1} \tag{7.32}$$

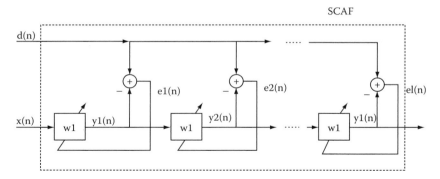

Figure 7.7

Book MATLAB function for self-correcting algorithm:
[w,y,e,J] = ssp_self_correcting_lms(x,dn,mu,M,I)

```
function[w,y,e,J]=ssp_self_correcting_lms(x,dn,mu,M,I)
%function[w,y,e,J]aaselfcorrectinglms(x,dn,mu,M,I);
[w(1,:),y(1,:),e(1,:),j(1,:)]=ssp_lms(x,dn,mu,M);
for i=2:I%I=number of iterations, I<8-10 is sufficient;
    [w(i,:),y(i,:),e(i,:),j(i,:)]=ssp_ ...
      lms(y(i-1,:),dn,mu,M);
end;
J=e.^2;
```

Figure 7.8 shows the results using the self-correcting algorithm with the following factors and constants:

$$s = 2\sin(0.1n\pi), \quad x(n) = s + \text{randn}, \quad M = 12, \quad I = 20, \quad \mu = 0.005$$

The output $y_1(n)$ is at the first stage and $y_{12}(n)$ is the output at the 12th stage.

Book MATLAB function for SCAF sign regressor:
[w,y,e,J] = ssp_self_correcting_sign_regressor_lms(x,dn,mu,M,I,ep)

```
function[w,y,e,J]=ssp_self_correcting_sign_regressor_ ...
  lms(x,dn,mu,M,I,ep)
%function[w,y,e,J]=ssp_self_correcting_sign_regressor_
%lms(x,dn,mu,M,I);
%x=input data to the filter;dn=desired
%signal;length(x)=length(dn);
```

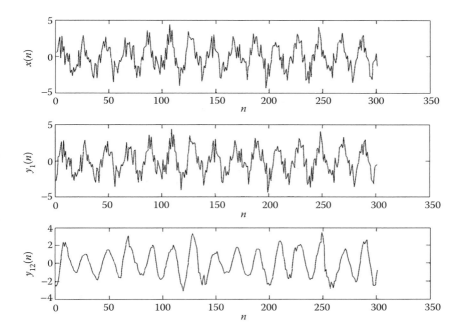

Figure 7.8

```
%y=output of the filter an Ix(length(x)) matrix;J=error
%function an
%Ix(length(x)) matrix;I=number of stages;ep≪1;
[w(1,:),y(1,:),e(1,:),J(1,:)]= ...
ssp_lms_norm_signed_regressor(x,dn,mu,M,ep);
for i=2:I
  [w(i,:),y(i,:),e(i,:),J(i,:)]= ...
  ssp_ lms(y(i-1,:),dn,mu,M);
end;
J=e.^2;
```

Figure 7.9 shows the results of a proposed self-correcting sign regressor algorithm. The constants and the factors are

$$s = \sin(0.1n\pi), \quad x = s + \text{randn}, \quad \mu = 0.01, \quad M = 10, \quad I = 10, \quad ep = 0.01$$

The outputs of the LMS filter are at the first and third stages are the learning curves for the same stages.

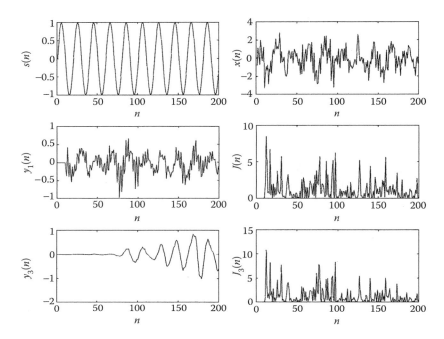

Figure 7.9

Book MATLAB function for self-correcting sign-sign algorithm:
[w,y,e,J] = ssp_self_correcting_sign_sign_lms(x,dn,mu,M,I)

```
function[w,y,e,J]=ssp_self_correcting_sign_sign_ ...
  lms(x,dn,mu,M,I)
%function[w,y,e,J]=ssp_self_correcting_sign_sign_
%lms(x,dn,mu,M,I);
%x=input data to the filter;y=output data from the
%filter,
%y is an Ixlength(x) matrix; J=learning curves, an
%Ixlength(x)
%matrix;mu=step-size parameter;M=umber of
%coefficients;I=
%number of stages;w=an Ixlength(x) matrix of filter
%coefficients;
%dn=desired signal;
[w(1,:),y(1,:),e(1,:),J(1,:)]= ...
ssp_lms_sign_sign(x,dn,mu,M);
```

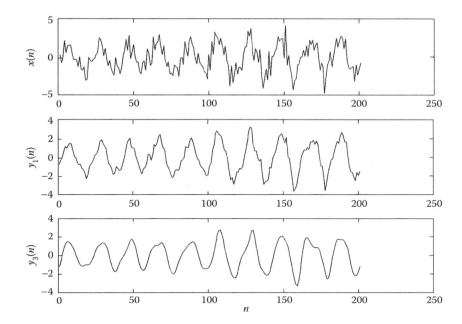

Figure 7.10

```
for i=2:I
    [w(i,:),y(i,:),e(i,:),J(i,:)]= ...
    ssp_ lms(y(i-1,:),dn,mu,M);
end;
J=e.^2;
```

Figure 7.10 presents the results using the above algorithm. The factors and constants used for this figure are

$$s(n) = 2\sin(0.1n\pi), \quad x(n) = s + \text{randn}, \quad d(n) = x(n-1),$$
$$\mu = 0.001, \quad M = 8, \quad I = 8.$$

The output of the filter at the first stage is $y_1(n)$ and the one from the third stage is $y_3(n)$.

7.7   Transform domain adaptive LMS filtering

The implementation of the LMS filter in the frequency domain can be accomplished simply by taking the discrete Fourier transform (DFT) of both the input data, $\{x(n)\}$, and the desired signal, $\{d(n)\}$. The advantage of doing this is

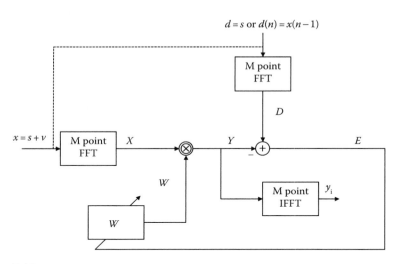

Figure 7.11

due to the fast processing of the signal using the fast Fourier transform (fft) algorithm. However, this procedure requires a block-processing strategy, which results in storing a number of incoming data in buffers, and thus some delay is unavoidable.

A simple approach has been suggested for this type of filtering and its diagrammatic form is shown in Figure 7.11. The signals are processed by a block-by-block format, that is $\{x(n)\}$ and $\{d(n)\}$ are sequenced into blocks of length M so that

$$x_i(n) = x(iM + n), \quad d_i(n) = d(iM + n), \quad n = 0, 1, \ldots, M-1, \quad i = 0, 1, \ldots \quad (7.33)$$

The values of the ith block of the signals $\{x_i(n)\}$ and $\{d_i(n)\}$ are Fourier transformed using the DFT to find $X_i(k)$ and $D_i(k)$, respectively. Due to DFT properties, the sequences have $X_i(k)$ and $D_i(k)$ have M complex elements corresponding to frequency indices ("bins") $k = 0, 1, \ldots, M - 1$

$$X_i(k) = \text{DFT}\{x_i(n)\} = \sum_{n=0}^{M-1} x_i(n)e^{-j(2\pi nk/M)}, \quad k = 0, 1, \ldots, M-1 \qquad (7.34)$$

$$D_i(k) = \text{DFT}\{d_i(n)\} = \sum_{n=0}^{M-1} d_i(n)e^{-j(2\pi nk/M)}, \quad k = 0, 1, \ldots, M-1 \qquad (7.35)$$

During the ith block processing, the output in each frequency bin of the adaptive filter is computed by

$$Y_i(k) = W_{i,k}X_i(k), \quad k = 0, 1, 2, \ldots, M-1 \tag{7.36}$$

where $W_{i,k}$ is the kth frequency bin corresponding to the ith update (corresponding to the ith block data). The error in the frequency domain is

$$E_i(k) = D_i(k) - Y_i(k), \quad k = 0, 1, 2, \ldots, M-1 \tag{7.37}$$

The system output is given by

$$y_i(n) \triangleq y(iM+n) = \text{IDFT}\{Y_i(k)\} = \frac{1}{M}\sum_{k=0}^{M-1} Y_i(k)e^{j(2\pi nk/M)}, \quad n = 0, 1, 2, \ldots, M-1 \tag{7.38}$$

To update the filter coefficients we use, by analogy to LMS recursion, the following recursion:

$$\mathbf{W}_{i+1} = \mathbf{W}_i + 2\mu\mathbf{E}_i \cdot \mathbf{X}_i^* \tag{7.39}$$

where

$$\mathbf{W}_{i+1} = \begin{bmatrix} W_{i+1,0} & W_{i+1,1} & \cdots & W_{i+1,M-1} \end{bmatrix}^{\mathrm{T}}$$

$$\mathbf{W}_i = \begin{bmatrix} W_{i,0} & W_{i,1} & \cdots & W_{i,M-1} \end{bmatrix}^{\mathrm{T}}$$

$$\mathbf{E}_i = \begin{bmatrix} E_i(0) & E_i(1) & \cdots & E_i(M-1) \end{bmatrix}^{\mathrm{T}}$$

$$\mathbf{X}_i^* = \begin{bmatrix} X_i^*(0) & X_i^*(1) & \cdots & X_i^*(M-1) \end{bmatrix}^{\mathrm{T}}$$

The dot "." in Equation 7.39 implies element-by-element multiplication and "*" stands for complex conjugate. If we assume X_i^* in the form:

$$\mathbf{X}_i = \text{diag}\{X_i(0) \quad X_i(1) \quad \cdots \quad X_i(M-1)\} = \begin{bmatrix} X_i(0) & 0 & \cdots & 0 \\ 0 & X_i(1) & \cdots & 0 \\ \vdots & \vdots & & \vdots \\ 0 & 0 & \cdots & X_i(M-1) \end{bmatrix} \tag{7.40}$$

then Equation 7.39 becomes

$$\mathbf{W}_{i+1} = \mathbf{W}_i + 2\mu\mathbf{X}_i^*\mathbf{E}_i \tag{7.41}$$

Therefore, Equations 7.33 through 7.39 constitute the frequency domain of the LMS algorithm.

The Book MATLAB function that gives the coefficients after *I* blocks (or iterations) is given below.

Book MATLAB Fourier transform LMS function: [A] = ssp_ft_lms(x,d,M,I,mu)

```
function[A]=ssp_ft_lms(x,d,M,I,mu)
%function[A]=ssp_ft_lms(x,d,M,I,mu);
wk=zeros(1,M);
for i=0:I     %I=number of iterations (or blocks);
  if I*M>length(x)-1
    ('error:I*M<length(x)-1')
  end;
            %M=number of filter coefficients;
            x1=x(M*(i+1):-1:i*M+1);
            d1=d(M*(i+1):-1:i*M+1);
  xk=fft(x1);
  dk=fft(d1);
  yk=wk.*xk;
  ek=dk-yk;
  wk=wk+2*mu*ek.*conj(xk);
  A(i+1,:)=wk;
end;
%all the rows of A are the wk's at an increase order
%of iterations(blocks);
%to filter the data, wk must be inverted in the time
%domain, convolve with the data x and then plot the
%real part of the output y, e.g. wn4=the forth iteration
%=ifft(A(4,:)),yn4=filter(wn4/4,1,x) for even M;
```

Figure 7.12 was created using the above algorithm in the following program:

```
M=16; I=200; mu=0.005;
n=0:2000;
s=sin(0.1*pi*n); v=randn(1,2000);
x=s+v;
for n=0:1999
dn(n+2)=x(n+1);
end;
[A]=ssp_ft_lms(x,dn,M,I,mu);
w2=ifft(A(2,:));% the inverse fft of row 2 of A;
y2=filter(w2/2,1,x);
w10=ifft(A(10,:));y10=filter(w10/10,1,x);
subplot(3,1,1);plot(x(1,1:200));ylabel('x(n)');
subplot(3,1,2);plot(real(y2(1,1:200)));ylabel('y₂(n)');
subplot(3,1,3);plot(real(y10(1,1:200)));xlabel('n');...
ylabel('y₁₀(n)');
```

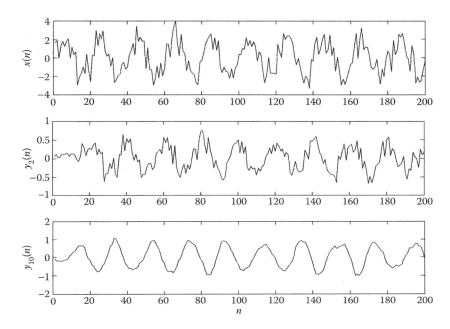

Figure 7.12

7.8 Convergence in transform domain of the adaptive LMS filtering

Let the signals $\{x(n)\}$ and $\{y(n)\}$ jointly stationary, and let the initial filter coefficients be zero, $\mathbf{W}_0 = 0$. Substituting Equations 7.36 and 7.37 in Equation 7.39 we obtain for the kth component the relation:

$$
\begin{aligned}
W_{i+1,k} &= W_{i,k} + 2\mu X_i^* E_i = W_{i,k} + 2\mu X_i^* \left[D_i(k) - Y_i(k) \right] \\
&= W_{i,k} + 2\mu X_i^*(k) \left[D_i(k) - W_{i,k} X_i(k) \right] \\
&= W_{i,k} + 2\mu X_i^*(k) D_i(k) - 2\mu W_{i,k} \mid X_i(k) \mid^2 \\
&= \left(1 - 2\mu \mid X_i(k) \mid^2 \right) W_{i,k} + 2\mu D_i(k) X_i^*(k)
\end{aligned}
\tag{7.42}
$$

The expected value of Equation 7.42, assuming $W_{i,k}$ and X_i are statistically independent, is given by

$$
E\{W_{i+1,k}\} = \left(1 - 2\mu E\left\{ \mid X_i(k) \mid^2 \right\}\right) E\{W_{i,k}\} + 2\mu E\left\{ D_i(k) X_i^*(k) \right\}
\tag{7.43}
$$

Because $x(n)$ and $x(k)$ are stationary, their statistical characteristics do not change from block to block and, therefore, the ensembles $E\{\mid X_i(k) \mid^2\}$ and $E\{D_i(k)X_i^*(k)\}$

are independent of i but depend on k. Taking the Z-transform of Equation 7.43 with respect to i of the dependent variable $W_{i,k}$, $Z\{W_{k,i}\} = W_k(z) = \sum_{i=0}^{\infty} W_{k,i} z^{-i}$) we obtain the following relation (the Z-transform and the ensemble are linear operations and can be interchanged):

$$z W_k(z) - z W_{0,k} = \left(1 - 2\mu E\{|X_i(k)|^2\}\right)W_k(z) + \left[\frac{2\mu E\{D_i(k)X_i^*(k)\}}{1 - z^{-1}}\right] \qquad (1)$$

where $W_{0,k} = 0$ since it was assumed that the initial conditions have zero values, and $W_k(z) = \sum_{i=0}^{\infty} E\{W_{i,k}\} z^{-i}$. Multiplying (1) by z^{-1} and $(z - 1)$, and applying the final value theorem ($\lim_{n \to \infty} x(n) = \lim_{n \to 1} (1 - z^{-1}) X(z)$) we obtain

$$E\{W_k^\infty\} = \lim_{z \to 1}\{z - 1\} W_k(z)\} = \lim_{z \to 1} \frac{2\mu E\{D_i(k)X_i^*(k)\}}{1 - (1 - 2\mu E\{|X_i(k)|^2\})z^{-1}} = \frac{E\{D_i(k)X_i^*(k)\}}{E\{|X_i(k)|^2\}} \qquad (7.44)$$

Let the mean filter coefficient error $E_i(k)$ be defined by

$$E_i(k) = E\{W_{i,k}\} - E\{W_k^\infty\} \qquad (7.45)$$

Then, using Equations 7.44 and 7.43 we find

$$E_{i+1}(k) = E\{W_{i+1,k}\} - E\{W_k^\infty\} = (1 - 2\mu E\{|X_i(k)|^2\})E\{W_{i,k}\} + 2\mu E\{D_i(k)X_i^*(k)\} - \frac{E\{D_i(k)X_i^*(k)\}}{E\{|X_i(k)|^2\}}$$

$$= (1 - 2\mu E\{|X_i(k)|^2\})E\{W_{i,k}\} - (1 - 2\mu E\{|X_i(k)|^2\})\frac{E\{D_i(k)X_i^*(k)\}}{E\{|X_i(k)|^2\}}$$

$$= (1 - 2\mu E\{|X_i(k)|^2\})E_i(k) \qquad (7.46)$$

Using the iteration approach (setting $i = 0$, $i = 1$, etc.) we find the solution of the above equation to be

$$E_i(k) = (1 - E\{|X_i(k)|^2\})^i E_0(k), \quad k = 0, 1, 2, \ldots, M-1 \qquad (7.47)$$

The solution converges if

$$\left|1 - 2\mu E\{|X_i(k)|^2\}\right| < 1 \quad \text{or} \quad 0 < \mu < \frac{1}{E\{|X_i(k)|^2\}} \qquad (7.48)$$

which shows that the power of the input plays a fundamental role in convergence and stability.

7.9   Error-NLMS algorithm

A new class of LMS algorithm based on error normalization has been proposed (Poularikas and Ramadan, 2006) and these are

1. **Error-normalized step-size** (ENSS) LMS algorithm

$$\mathbf{w}(n+1) = \mathbf{w}(n)\frac{\mu}{1+\mu \|\mathbf{e}_{L(n)}\|^2}\mathbf{x}(n)e(n) \tag{7.49}$$

2. **Robust variable step-size** (RVSS) LMS algorithm

$$\mathbf{w}(n+1) = \mathbf{w}(n)+\frac{\mu \|\mathbf{e}_L(n)\|^2}{\alpha \|\mathbf{e}(n)\|^2 + (1-\alpha)\|\mathbf{x}(n)\|^2}\mathbf{x}(n)e(n) \tag{7.50}$$

3. **Error-data-normalized step-size** (EDNSS) LMS algorithm

$$\boxed{\mathbf{w}(n+1) = \mathbf{w}(n)+\frac{\mu}{\alpha \|\mathbf{e}_L(n)\|^2 + (1-\alpha)\|\mathbf{x}(n)\|^2}\mathbf{x}(n)e(n)} \tag{7.51}$$

where

$$\|\mathbf{e}_L(n)\|^2 = \sum_{i=0}^{L-1} |e(n-i)|^2 \tag{7.52}$$

and

$$\|\mathbf{e}(n)\|^2 = \sum_{i=0}^{n-1} |e(n-i)|^2 \tag{7.53}$$

Comments

- The parameters α, L, and μ in all of these algorithms are appropriately chosen to achieve the best trade-off between rate of convergence and low final MSE. L could be constant or variable ($L = n$, for example), depending on whether the underlying environment is stationary or nonstationary.
- The variable step-sizes in all of these algorithms should vary between two predetermined hard limits. The lower value guarantees the capability of the algorithm to respond to any abrupt change that could happen at a very large value of iteration number n, while the maximum value maintains stability of the algorithm.

Book MATLAB function for the ENSS algorithm:
[w,y,e,J,dd,n] = ssp_error_normalized_step_size(I,LL,m1)

```
function[w,y,e,J,dd,n]=ssp_error_normalized_step_ ...
   size(I,LL,mu1)
```

```
%NN=length of the error vector e sub L;mu1=step size;
%N=length of adaptive filter;
%I=number of independent simulations runs used to
%average
%the learning curv;
%LL=total number of iterations;
%h=the impulse response of an unknown system(plant);
%x=input to the unknown system and adaptive filter;
%dd=output of the unknown plant used as the desired
%signal;
%y=output of the adaptive filter;
%n=the internal noise of the unknown system;
%J=the MSE;

J=zeros(1,LL);Jminn=zeros(1,LL);
N=4;NN=10*N; h=[1 0.72 0.5 -0.21];
J=zeros(1,LL);Jex=zeros(1,LL);
for i=1:I
  X=zeros(N,1);D=zeros(NN,1);y=zeros(1,LL);
  x=randn(1,LL);w=zeros(1,N);e=zeros(1,LL);
  denn=0;n=sqrt(0.09)*(randn(1,LL));

  for k=1:LL
      dd=filter(h,1,x);
      X=[x(k);X(1:N-1)];den=X'*X;y(k)=w*X;
      e(k)=dd(k)+n(k)-y(k);
      denn=denn+e(k)^2;mu=(mu1/(1+mu1*denn));
      w=w+mu*e(k)*X';
      J(k)=J(k)+(abs(e(k)))^2;
      Jminn(k)=Jminn(k)+n(k)^2;
  end;
end;
J=J/I;Jmin1=Jminn/I;Jmin=sum(Jmin1)/LL;Jex=J-Jmin;
Jinf=(1/I)*sum(J(LL-I-1:LL));JSSdB=10*log10(Jinf);
Jexinf=abs(Jinf-Jmin);JexinfSSdB=10*log10(Jexinf);
MM=Jexinf/Jmin;Mpercent=MM*100;
```

Book MATLAB function for average LMS filter corresponding to the above algorithm:
function[w,y,J,Mpercent] = ssp_average_normalized_lms(I,LL,mub,ep)

```
function [w,y,J,Mpercent]=ssp_average_normalized_ ...
  lms(I,LL,mub,ep)
%N=length of adaptive filter;
%I=number of independent simulations runs used to
%average
```

```
%the learning curv;
%LL=total number of iterations;
%h=the impulse response of an unknown system(plant);
%x=input to the unknown system and adaptive filter;
%dd=output of the unknown plant used as the desired
%signal;
%y=output of the adaptive filter;
%n=the internal noise of the unknown system;
%J=the MSE;ep<<1;mub=mu bar;
J=zeros(1,LL);Jminn=zeros(1,LL);
N=4; h=[1 0.72 0.5 -0.21];
J=zeros(1,LL);Jex=zeros(1,LL);
for i=1:I
    X=zeros(N,1);y=zeros(1,LL);
    x=randn(1,LL);w=zeros(1,N);e=zeros(1,LL);
    n=sqrt(0.09)*(randn(1,LL));

  for k=1:LL
    dd=filter(h,1,x);
    X=[x(k);X(1:N-1)];den=X'*X+ep;y(k)=w*X;
    e(k)=dd(k)+n(k)-y(k);

    w=w+(mub*e(k)*X'/(den));
    J(k)=J(k)+(abs(e(k)))^2;
    Jminn(k)=Jminn(k)+n(k)^2;
  end;
end;
J=J/I;Jmin1=Jminn/I;Jmin=sum(Jmin1)/LL;Jex=J-Jmin;
Jinf=(1/I)*sum(J(LL-I-1:LL));JSSdB=10*log10(Jinf);
Jexinf=abs(Jinf-Jmin);JexinfSSdB=10*log10(Jexinf);
MM=Jexinf/Jmin;Mpercent=MM*100;
```

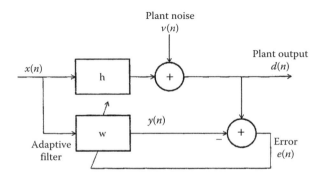

Figure 7.13

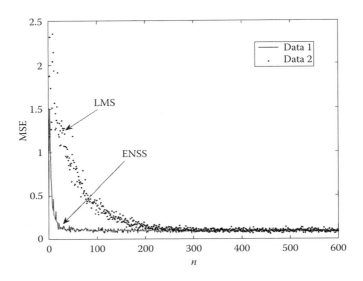

Figure 7.14

Simulations

The proposed ENSS algorithm using an error vector of increasing length ($L = n$) is compared with the LMS algorithm in system identification as in Figure 7.13. The adaptive filter noise and the unknown system are both excited by a zero mean white Gaussian signal with unit variance. The length of the unknown filter is assumed to be of the fourth order ($N = 4$). The internal noise, $v(n)$, of the unknown system is assumed to be white Gaussian with zero mean value and variance $\sigma_v^2 = 0.09$. The optimum value of μ, in both algorithms, is chosen to obtain approximately the same mis-adjustment value, $M = 4\%$. Figure 7.14 shows the results which indicate that the proposed ENSS filter is superior to LMS one. The curves are the result of 100 independent simulations.

chapter eight

Nonlinear filtering

8.1 Introduction

Nonlinear filtering techniques remove unknown interference to one-dimensional signals $f(n)$ or two-dimensional signals (images) $f(m, n)$, where the integers m and n have, in practice, finite images: $0 \le m \le M - 1$, $0 \le n \le N - 1$. The basic problem is to use the received (or detected) signal $\mathbf{x} = [x_0\, x_1 \cdots x_{N-1}]^N$, which is corrupted by noise, and try to extract the original signal $\mathbf{s} = [s_0\, s_1 \cdots s_{N-1}]$, which is another random vector. Although it is rather difficult to find a useful solution, by making simplifications we may solve the problem with the available methods at hand. Complexity of the reduced solution depends on (1) the model of the underlying signal $\mathbf{s}$, (2) the nature of the additive noise (corruption) to the signal, and (3) the accuracy of the solution with respect to the above assumptions.

It is known that when the noise is Gaussian process and linearly added to the signal, the theory of the linear filtering gives optimum solutions. In this case, the mean square error is used as the criterion of accuracy.

There are cases, and this is more pronounced in images, that the intensity of a pixel is due to the signal (desired image) plus scattering light from different parts of the environment during its formation. In such a situation, the white noise additive model seldom holds. The intensity of each pixel of an image created by an acquisition system is usually a **multiplicative** process with respect to background illumination. This is equivalent to say that the ith pixel intensity is equal $s_i v_i$, where s_i is the signal intensity and v_i is the noise intensity. Another important noise is the **impulsive noise**, which is recognized if the pixel values do not change at all or change slightly and some pixel values change enormously, i.e., their change is highly visible. Because this type of noise is hard to handle with linear filters, the use of **nonlinear filters** is more appropriate. In this chapter, we shall deal only with order-based filters, which have been employed successfully to pass the desired signal (image) structures while suppressing noise.

8.2 Statistical preliminaries

8.2.1 Signal and noise model—robustness

The simplest and most important case of signal and noise is the additive white noise model where the signal $\mathbf{s}$ and noise $\mathbf{v}$ are assumed independent.

The white indicates that v_is are independent or at most uncorrelated. For this model we write

$$\mathbf{x} = \mathbf{s} + \mathbf{v} \tag{8.1}$$

As we mentioned in the previous section, the multiplicative noise model is expressed as

$$\mathbf{x} = \mathbf{sv} \tag{8.2}$$

or in expanded form:

$$[x_0 \quad x_1 \quad \cdots \quad x_{N-1}] = [s_0 v_0 \quad s_1 v_1 \quad \cdots \quad s_{N-1} v_{N-1}] \tag{8.3}$$

where s_is are independent of v_is.

Another type of noise is the impulsive noise, also known as **outliers**. An outlier can be defined as an observation which appears to be inconsistent with the remaining data.

8.2.2   Point estimation

In a typical case, we are faced with the problem of extracting **parameter** values from a discrete-time waveform or a data set. For example, let us have an N-point data set $\{x(0) \; x(1) \cdots x(N-1)\}$ which depends on an unknown parameter θ. Our aim is to determine θ based on the data or define an **estimator**

$$\hat{\theta} = g(x(0), x(1), \ldots, x(N-1)) \tag{8.4}$$

where g is some function (statistic) of the data, and its numerical value is called an estimate of θ.

To determine a good estimator, we must model the data mathematically. Because the data are inherently random, we describe them by the probability density function (pdf) or $p(x(0), x(1), \ldots, x(N-1); \theta)$ which is **parameterized** by the unknown parameter θ. This type of dependence is denoted by semicolon.

As an example, let $x(0)$ be a random variables from a Gaussian population and with mean value $\theta = \mu$. Hence, the pdf is

$$p(x(0); \theta) = \frac{1}{\sqrt{2\pi\sigma^2}} e^{-(1/2\sigma^2)(x(0)-\theta)^2} \tag{8.5}$$

The plots of $p(x(0); \theta)$ for different values of θ are shown in Figure 8.1. From the figure, we can **infer** that, if the value of $x(0)$ is positive, it is doubtful that

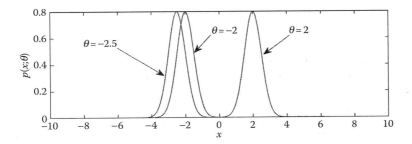

Figure 8.1

$\theta = \theta_2$ or θ_3 and, hence, the value $\theta = \theta_1$ is more probable. In the area of point estimation, the specification of the appropriate pdf is critical in determining a good estimator. In the case when the pdf is not given, which is more often than not, we must try to choose one that is consistent with the constraints of any prior knowledge of the problem, and furthermore that is mathematically tractable.

Once the pdf has been selected, it is our concern to determine an optimal estimator (a function of the data). The estimator may be thought of as a rule that assigns a value to θ for each realization of the sequence $\{x\}_N$. The estimate of θ is the **value** of θ obtained at a particular realization $\{x\}_N$.

8.2.3   Estimator performance

Let the set $\{x(0), x(1), \ldots, x(N-1)\}$ of random data is the sum of a constant c and a zero mean white noise $v(n)$

$$x(n) = c + v(n) \tag{8.6}$$

Intuitively, we may set as an estimate of c the sample mean of the data

$$\hat{c} = \frac{1}{N}\sum_{n=0}^{N-1} x(n) \tag{8.7}$$

From Figure 8.2, we find that $x(0) = 0.2837$ and may accept the random variable $x(0)$ as another estimate of the mean

$$\tilde{c} = x(0) \tag{8.8}$$

The basic question is: which of these two estimators will produce the more accurate mean value. Instead of repeating the experiment a large number of

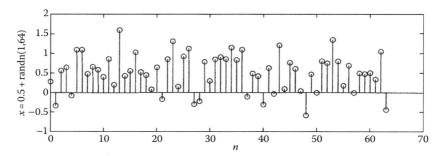

Figure 8.2

times, we proceed to prove that sample mean is a better estimator than $x(0)$. To do this, we first look at their mean value (expectation):

$$E\{\hat{c}\} = E\left\{\frac{1}{N}\sum_{n=0}^{N-1}x(n)\right\} = \frac{1}{N}\sum_{n=0}^{N-1}E\{x(n)\}$$

$$= \frac{1}{N}E\{x(0) + x(1) + \cdots + x(N-1)\} = \frac{N\mu}{N} = \mu \qquad (8.9)$$

$$E\{\tilde{c}\} = E\{x(0)\} = \mu \qquad (8.10)$$

which indicates that on the average both estimators produce the true mean value of the population. Next, we investigate their variances which are (see Problem 8.2.1):

$$\text{var}\{\tilde{c}\} = \text{var}\left\{\frac{1}{N}\sum_{n=0}^{N-1}x(n)\right\} = \frac{\sigma^2}{N} \qquad (8.11)$$

$$\text{var}\{\tilde{c}\} = \text{var}\{x(0)\} = \sigma^2 \qquad (8.12)$$

The above results show that $\text{var}\{\tilde{c}\} > \text{var}\{\hat{c}\}$, and the $\text{var}\{\hat{c}\}$ approaches 0 as $N \to \infty$. To prove Equation 8.11, we assume that $v(n)$s are iid and have the same variance σ^2.

8.2.4   Biased and unbiased estimator

An estimator which **on the average** yields the true value of the unknown parameter is known as unbiased one and, mathematically, is given by

$$E\{\hat{\theta}\} = \theta, \quad a < \theta < b \qquad (8.13)$$

where (a, b) denotes the range of possible values of θ (see Problem 8.2.2).

An unbiased estimator is given mathematically by the relation (see Problem 8.2.3):

$$E\{\hat{\theta}\} = \theta + b(\theta) \tag{8.14}$$

where the bias of the estimator is given by

$$b(\theta) = E\{\hat{\theta}\} - \theta \tag{8.15}$$

8.2.5   Cramer–Rao lower bound (CRLB)

It is helpful to place a lower bound on the variance of any unbiased estimator. The Cramer–Rao bound is the appropriate measure. It assures us if the estimator is the minimum variance unbiased estimator (MVUE) estimator or provide us with a benchmark to compare the performance of the estimator.

Theorem 8.1 (Cramer–Rao lower bound)   *It is assumed that the pdf $p(\mathbf{x};\theta)$ satisfies the regularity condition*

$$E\left\{\frac{\partial \ln p(\mathbf{x};\theta)}{\partial \theta}\right\} = 0, \quad \text{for all } \theta \tag{8.16}$$

where the expectation is taken with respect to $p(\mathbf{x};\theta)$. Then, the variance of any unbiased estimator $\hat{\theta}$ must satisfy

$$\boxed{\text{var}\{\hat{\theta}\} \geq \frac{1}{-E\left\{(\partial^2 \ln p(\mathbf{x};\theta))/\partial\theta^2\right\}}}$$

$$E\left\{\frac{\partial^2 \ln p(\mathbf{x};\theta)}{\partial\theta^2}\right\} = \int \frac{\partial^2 \ln p(\mathbf{x};\theta)}{\partial\theta^2} p(\mathbf{x};\theta)\,d\mathbf{x} \tag{8.17}$$

where the derivative is evaluated at the true value of θ and the expectation is taken with respect to $p(\mathbf{x};\theta)$. Furthermore, an unbiased estimator may be found that contains the bound for all θ if and only if

$$\frac{\partial \ln p(\mathbf{x};\theta)}{\partial \theta} = I(\theta)(g(\mathbf{x}) - \theta) \tag{8.18}$$

for some function of g and I. That estimator, which is the MVUE, is $\hat{\theta} = g(\mathbf{x})$, and the minimum variance (MV) is $1/I(\theta)$.   ∎

Let us consider a number of observations

$$x(n) = c + v(n), \quad n = 0,1,\ldots,N-1 \tag{8.19}$$

where $v(n)$ is a white Gaussian noise (WGN) with variance σ^2. To determine the CRLB for c (vs are iid), we find the pdf

$$p(\mathbf{x};c) = \prod_{n=0}^{N-1} \frac{1}{(2\pi\sigma^2)^{1/2}} \exp\left(-\frac{1}{2\sigma^2}(x(n)-c)^2\right)$$

$$= \frac{1}{(2\pi\sigma^2)^{N/2}} \exp\left(-\frac{1}{2\sigma^2}\sum_{n=0}^{N-1}(x(n)-c)^2\right) \qquad (8.20)$$

Taking the first derivative, we obtain

$$\frac{\partial \ln p(x;\theta)}{\partial c} = \frac{\partial}{\partial c}\left(-\ln[(2\pi\sigma^2)^{N/2}] - \frac{1}{2\sigma^2}\sum_{n=0}^{N-1}(x(n)-c)^2\right)$$

$$= \frac{1}{\sigma^2}\sum_{n=0}^{N-1}(x(n)-c) = \frac{N}{\sigma^2}(\overline{x}-c), \quad \overline{x} = \text{sample mean} \qquad (8.21)$$

Differentiating once again, we obtain

$$\frac{\partial^2 \ln p(\mathbf{x};\theta)}{\partial c^2} = -\frac{N}{\sigma^2} \qquad (8.22)$$

Since the second derivative is a constant, from Equation 8.17, we obtain the CRLB

$$\text{var}(\hat{c}) \geq \frac{1}{N/\sigma^2} = \frac{\sigma^2}{N} \qquad (8.23)$$

Comparing Equation 8.21 with Equation 8.18, we find the following correspondent relations:

$$I = \frac{N}{\sigma^2} \quad \text{and} \quad (g(\mathbf{x})-\theta) = (\overline{x}-c) \qquad (8.24)$$

These relations give us the MV and the MVUE.

8.2.6   Mean square error criterion

When we seek the minimum mean square estimator, we can use a simplified form of the estimator where the statistic is a linear combination of the random data set $\{x(0), x(1), \ldots, x(N-1)\}$. Hence, we need to determine $a_0, a_1, \ldots, a_{N-1}$ such that

$$\text{mse}(\hat{\theta}) = E\{(\hat{\theta}-\theta)^2\} = E\{[a_0 x(0) + a_1 x(1) + \cdots + a_{N-1} x(N-1)]^2\} \qquad (8.25)$$

is minimized. Since expectation is a linear operation, the above equation becomes

$$\text{mse}(\hat{\theta}) = E\{\theta^2\} - 2\sum_{n=0}^{N-1} a_i E\{\theta x(n)\} + \sum_{m=0}^{N-1}\sum_{n=0}^{N-1} a_m a_n E\{x(m)x(n)\} \qquad (8.26)$$

The above equation is quadratic in a_ns and setting the derivatives equal to 0, we obtain the following set of equations (see Problem 8.2.4):

$$r_{00}a_0 + r_{01}a_1 + \cdots + r_{0N-1}a_{N-1} = r(\theta, x(0)) \triangleq r_{\theta x}(0)$$
$$r_{10}a_0 + r_{11}a_1 + \cdots + r_{1N-1}a_{N-1} = r(\theta, x(1)) \triangleq r_{\theta x}(1)$$
$$\vdots$$
$$r_{N-1,0}a_0 + r_{N-1,1}a_1 + \cdots + r_{N-1,N-1}a_{N-1} = r(\theta, x(N-1)) \triangleq r_{\theta x}(N-1)$$

or

$$\mathbf{R a} = \mathbf{r}_{\theta x} \qquad (8.27)$$

where
$$r_{mn} = E\{x(m)x(n)\}$$
$$r_{\theta x}(m) = E\{\theta x(m)\}$$

From Equation 8.25, we can also proceed as follows:

$$\mathrm{mse}(\hat{\theta}) = E\{[(\hat{\theta} - E\{\hat{\theta}\}) + (E\{\hat{\theta}\} - \theta)^2\}$$
$$= E\{[(\hat{\theta} - E\{\hat{\theta}\}]^2\} + E\{[E\{\hat{\theta}\} - \theta]^2\} + 2E\{[\hat{\theta} - E\{\hat{\theta}\}][E\{\hat{\theta}\} - \theta]\}$$
$$= \mathrm{var}(\hat{\theta}) + [E\{\hat{\theta}\} - \theta]^2 = \mathrm{var}(\hat{\theta}) + b^2(\theta) \qquad (8.28)$$

which shows that the mean square error is the sum of the error due to the variance of the estimator as well as its bias. Constraining the bias to 0, we can find the estimator which minimizes its variance. Such an estimator is known as the **minimum variance unbiased estimator** (MVUE).

8.2.7 *Maximum likelihood estimators*

Very often, MVU estimators may be difficult or impossible to determine. For this reason, in practice, many estimators are found using the **maximum likelihood estimator** (MLE) principle. Besides being easy to implement, its performance is optimal for large number of data. The basic idea is to find a statistic

$$\hat{\theta} = g(x(0), x(1), \ldots, x(N-1)) \qquad (8.29)$$

so that if the random variables $x(m)$s take the observed experimental value $x(m)$s, then the number $\hat{\theta} = g(x(0), x(1), \ldots, x(N-1))$ will be a good estimate of θ.

Definition 8.1   Let $x = [x(0)\ x(1) \ldots x(N-1)]^T$ be a random vector with density function:

$$p(x(0), x(1), \ldots, x(N-1); \theta), \quad \theta \in \Theta$$

The function

$$l(\theta; x(0), x(1), \ldots, x(N-1)) = p(x(0), x(1), \ldots, x(N-1); \theta) \quad (8.30)$$

is considered as a function of the parameter $\boldsymbol{\theta} = [\theta_0\ \theta_1 \cdots \theta_{N-1}]^{\mathrm{T}}$, is called the **likelihood function** (*l* identifies the likelihood function with one parameter (scalar)). ∎

The random variables $x(0)$, $x(1)$, ..., $x(N-1)$ are iid with a density function $p(\mathbf{x}; \boldsymbol{\theta})$, then the likelihood function is

$$L(\theta; x(0), x(1), \ldots, x(N-1)) = \prod_{n=0}^{N-1} p(x(n); \theta) \quad (8.31)$$

Example 8.1

Let $x(0)$, $x(1)$, ..., $x(N-1)$ be random sample from the normal distribution $N(\theta, 1)$, $-\infty < \theta < \infty$ ($\theta \triangleq \mu$ = mean of the population). Using Equation 8.31, we write

$$L(\theta; x(0), x(1), \ldots, x(N-1)) = \left(\frac{1}{\sqrt{2\pi}}\right)^N \exp\left(-\sum_{n=0}^{N-1} \left((x(n)-\theta)^2/2\right)\right) \quad (8.32)$$

Since the likelihood function $L(\theta)$ and its logarithm $\ln\{L(\theta)\}$ are maximized for the same value of the parameter θ, we can use either $L(\theta)$ or $\ln\{L(\theta)\}$. Therefore,

$$\frac{\partial \ln\{L(\theta; x(0), x(1), \ldots, x(N-1))\}}{\partial \theta} = \frac{\partial}{\partial \theta}\left(N\ln\{1/\sqrt{2\pi}\} - \sum_{n=0}^{N-1}\left((x(n)-\theta)^2/2\right)\right)$$

$$= -\sum_{n=0}^{N-1}(x(n)-\theta) = 0 \quad (8.33)$$

and the solution of Equation 8.33 for θ is

$$\hat{\theta} = g(x(0), x(1), \ldots, x(N-1)) = \frac{1}{N}\sum_{n=0}^{N-1}x(n) \quad (8.34)$$

The above equation shows that the estimator maximizes $L(\theta)$. Therefore, the statistic $g(\cdot)$ above (the sample mean value of the data) is the maximum likelihood estimator of the mean, $\theta \triangleq \mu$.

Since $E\{\hat{\theta}\} = (1/N)\sum_{n=0}^{N-1}E\{x(n)\} = N\theta/N = \theta$, the estimator is an unbiased one. ∎

Definition 8.2   If we choose a function $g(\mathbf{x}) = g(x(0), x(1), \ldots, x(N-1))$ such that θ is replaced by $g(\mathbf{x})$, the likelihood function L is maximum. That is, $L(g(\mathbf{x}); x(0), \ldots, x(N-1))$ is at least as great as $L(\theta; x(0), x(1), \ldots, x(N-1))$ for all $\theta \in \Theta$, or in mathematical form:

$$L\left(\hat{\theta}; x(0), x(1), \ldots, x(N-1)\right) = \sup_{\theta \in \Theta} L\left(\theta; x(0), x(1), \ldots, x(N-1)\right) \quad (8.35)$$

∎

Definition 8.3   Any statistic whose expectation is equal to a parameter θ is called an unbiased estimator of the parameter θ. Otherwise, the statistic is said to be biased.

$$E\{\hat{\theta}\} = \theta, \quad \text{unbiased estimator} \quad (8.36)$$

∎

Definition 8.4   Any statistic that converges stochastically to a parameter θ is called a consistent estimator of that parameter. Mathematically, we write $\lim_{N \to \infty} \Pr\{|\hat{\theta} - \theta| > \varepsilon\} = 0.$

∎

If as $N \to \infty$ the relation:

$$\hat{\theta} \to \theta$$

holds, the estimator $\hat{\theta}$ is said to be a **consistent estimator**. If, in addition, as $N \to \infty$ the relation

$$E\{\hat{\theta}\} \to \theta$$

holds is said that the estimator $\hat{\theta}$ is said to be **asymptotically unbiased**. Furthermore, if as $N \to \infty$ the relation

$$\text{var}\{\hat{\theta}\} \to \text{lowest value}, \quad \text{for all } \theta$$

holds, then it is said that $\hat{\theta}$ is **asymptotically efficient**.

Example 8.2

Let the observed data $x(n)$ be the set

$$x(n) = c + v(n), \quad n = 0, 1, \ldots, N-1 \quad (8.37)$$

where c is an unknown constant greater than 0, and $v(n)$ is WGN with zero mean and with unknown variance c. The pdf is

$$p(\mathbf{x};c) = \frac{1}{(2\pi c)^{N/2}} \exp\left[-\frac{1}{2c}\sum_{n=0}^{N-1}(x(n)-c)^2\right] \qquad (8.38)$$

Considering Equation 8.38 as a function of c, it becomes a likelihood function $L(c\text{:}\mathbf{x})$. Differentiating its natural logarithm, we obtain

$$\frac{\partial \ln[(2\pi)^{N/2}p(\mathbf{x};c)]}{\partial c} = \frac{\partial}{\partial c}\left[-\frac{N}{2}\ln c - \frac{1}{2c}\sum_{n=0}^{N-1}(x(n)-c)^2\right]$$

$$= -\frac{N}{2}\frac{1}{c} + \frac{1}{c}\sum_{n=0}^{N-1}(x(n)-c) + \frac{1}{2c^2}\sum_{n=0}^{N-1}(x(n)-c)^2 = 0 \qquad (8.39)$$

where a multiplication of the pdf by a constant does not change the maximum point. From Equation 8.39, we obtain

$$\hat{c}^2 + \hat{c} - \frac{1}{N}\sum_{n=0}^{N-1}x^2(n) = 0$$

Solving for $\hat{c}$ and keeping the positive sign of the quadratic root, we find

$$\hat{c} = -\frac{1}{2} + \sqrt{\frac{1}{N}\sum_{n=0}^{N-1}x^2(n) + \frac{1}{4}} \qquad (8.40)$$

Note $\hat{c} > 0$ for all values of the summation under the square root. Since

$$E\{\hat{c}\} = E\left\{-\frac{1}{2} + \sqrt{\frac{1}{N}\sum_{n=0}^{N-1}x^2(n) + \frac{1}{4}}\right\} \neq -\frac{1}{2} + \sqrt{E\left\{\frac{1}{N}\sum_{n=0}^{N-1}x^2(n) + \frac{1}{4}\right\}} \qquad (8.41)$$

implies that the estimator is biased. From the law of the large numbers as $N \to \infty$

$$\frac{1}{N}\sum_{n=0}^{N-1}x^2(n) \to E\left\{x^{2(n)}\right\} = \mathrm{var}(x) + E\{x^2\} = c + c^2,$$

$$E\{(x-\bar{x})^2\} \triangleq \mathrm{var}(x) = E\{x^2\} - 2E\{x\}\bar{x} + \bar{x}^2 = E\{x^2\} - \bar{x}^2 \qquad (8.42)$$

and, therefore, Equation 8.40 gives

$$\left(\hat{c} + \frac{1}{2}\right)^2 = \left(c + c^2 + \frac{1}{4}\right) \quad \text{or} \quad \bar{c}^2 + \frac{1}{4} + \bar{c} = c + c^2 + \frac{1}{4}$$

which indicates that

$$\bar{c} \to c \tag{8.43}$$

And, hence, the estimator is a consistent estimator. ■

Example 8.3

It is required to find the maximum likelihood estimator for the mean m and variance σ^2 of a set of data $\{x(n)\}$ provided by a Gaussian random generator.

The Gaussian pdf $p(x; m, \sigma^2)$ for one rv is

$$p(x; m, \sigma^2) = (2\pi\sigma^2)^{-1/2} \exp\left[-\frac{1}{2}\left(\frac{x-m}{\sigma}\right)^2\right] \tag{8.44}$$

Its natural logarithm is

$$\ln p(x; m, \sigma^2) = -\frac{1}{2}\ln(2\pi\sigma^2) - \frac{1}{2}\left(\frac{x-m}{\sigma}\right)^2 \tag{8.45}$$

The likelihood function for the data (iid) is given by

$$L(m, \sigma^2) = p(x(0); m, \sigma^2)p(x(1); m, \sigma^2) \cdots p(x(N-1); m, \sigma^2) \tag{8.46}$$

and its logarithm is

$$\ln L(m, \sigma^2) = \sum_{n=0}^{N-1} \ln p(x(n); m, \sigma^2) \tag{8.47}$$

Substituting Equation 8.45 in Equation 8.47, we obtain

$$\ln L(m, \sigma^2) = \sum_{n=0}^{N-1}\left[-\frac{1}{2}\ln(2\pi\sigma^2) - \frac{1}{2}\left(\frac{x(n)-m}{\sigma}\right)^2\right]$$

$$= -\frac{N}{2}\ln(2\pi\sigma^2) - \frac{1}{2\sigma^2}\sum_{n=0}^{N-1}(x(n)-m)^2 \tag{8.48}$$

There are two unknowns in the log of the likelihood function. Differentiating Equation 8.48 with respect to the mean and variance, we obtain

$$\frac{\partial \ln L}{\partial m} = \frac{1}{\sigma^2} \sum_{n=0}^{N-1} (x(n) - m), \quad \frac{\partial \ln L}{\partial (\sigma^2)} = -\frac{N}{2} \frac{1}{\sigma^2} + \frac{1}{2\sigma^4} \sum_{n=0}^{N-1} (x(n) - m)^2 \quad (8.49)$$

Equating the partial derivatives to 0, we obtain

$$\frac{1}{\hat{\sigma}^2} \sum_{n=0}^{N-1} (x(n) - \hat{m}) = 0 \tag{8.50}$$

$$-\frac{N}{2} \frac{1}{\hat{\sigma}^2} + \frac{1}{2\hat{\sigma}^4} \sum_{n=0}^{N-1} (x(n) - \hat{m})^2 = 0 \tag{8.51}$$

For the estimate variance to be different than 0, Equation 8.50 reduces to

$$\sum_{n=0}^{N-1} (x(n) - \hat{m}) = 0 \quad \text{or} \quad \hat{m} = \frac{1}{N} \sum_{n=0}^{N-1} x(n) = \text{sample mean} \tag{8.52}$$

Therefore, the MLE of the mean of a Gaussian population is equal to the sample mean, which indicates that the sample mean is an optimal estimator.

Multiplying Equation 8.51 by $2\hat{\sigma}^4$ leads to

$$-N\hat{\sigma}^2 + \sum_{n=0}^{N-1} (x(n) - \hat{m})^2 = 0 \quad \text{or} \quad \hat{\sigma}^2 = \frac{1}{N} \sum_{n=0}^{N-1} (x(n) - \hat{m})^2 \tag{8.53}$$

Therefore, the MLE of the variance of a Gaussian population is equal to the sample variance. ∎

Let the pdf of the population be the Laplacian (a is a positive constant):

$$p(x; \theta) = \frac{a}{2} \exp(-a |x - \theta|) \tag{8.54}$$

Then, the likelihood function corresponding to the above pdf for a set of data $\{x(n)\}$(iid) is

$$L(\theta) = \ln\left\{\left(\frac{a}{2}\right)^N \prod_{n=0}^{N-1} \exp\left(-a\left|x(n)-\theta\right|\right)\right\}$$

$$= -a\sum_{n=0}^{N-1}\left|x(n)-\theta\right| + N\ln\{a\} - N\ln\{2\} \qquad (8.55)$$

Before we proceed further, we must define the term median for a set of random variables. Median is the value of a rv of the set when half of the rvs of the set have values less than the median and half have higher values (odd number of terms). Hence, the

$$\text{med}\{1,\ 4,\ 2,\ 3,\ 5,\ 9,\ 8,\ 7,\ 11\} = \text{med}\{1,\ 2,\ 3,\ 4,\ 5,\ 7,\ 8,\ 9,\ 11\} = 5$$

Definition 8.5 Let us consider a rv whose cdf (distribution) is F_x. The point x_{med} is the median of x if

$$F_x(x_{med}) = \frac{1}{2} \qquad (8.56)$$

∎

From Equation 8.55, we observe that the derivative of $-\sum_{n=0}^{N-1}\left|x(n)-\theta\right|$ with respect to θ is negative if θ is larger than the sample median, and positive if it is less than the sample median (remember that $\left|x(n)-\theta\right| = [(x(n)-\theta)(x(n)-\theta)]^{1/2}$ for real values). Therefore, the estimator is (see Problem 8.2.8):

$$\hat{\theta} = \text{med}\{x(0)\quad x(1)\quad \cdots \quad x(N-1)\} \qquad (8.57)$$

which maximizes $L(\theta)$ for the Laplacian pdf likelihood function.

Note: (1) From the above discussion we have observed that the minimization of $L(\theta)$ created from a Gaussian distribution is equivalent in minimizing $\sum_{n=0}^{N-1}(x(n)-\theta)^2$. The minimization results in finding the estimator $\hat{\theta} = (1/N)\sum_{n=0}^{N-1}x(n)$ which is the sample mean of the data $x(n)$. (2) Similarly, the minimization of the likelihood function $L(\theta)$ created from a Laplacian distribution is equivalent in minimizing $\sum_{n=0}^{N-1}\left|x(n)-\theta\right|$. This minimization results in finding the estimator:

$$\hat{\theta} = \text{med}\{x(0)\quad x(1)\quad \cdots \quad x(N-1)\}$$

8.3 Mean filter

Let us consider the following signal:

$$x(n) = s(n) + v(n), \quad n = 0, 1, \ldots, N-1 \qquad (8.58)$$

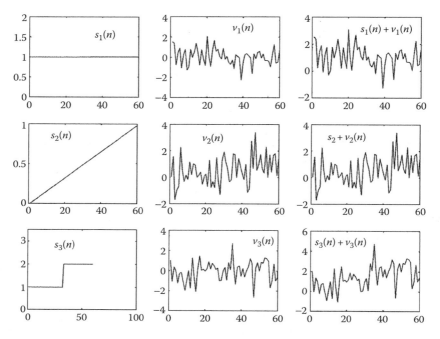

Figure 8.3

where $v(n)$ is WGN with zero mean and variance about 1. For the determin-istic signal, we have selected (1) a constant, (2) a ramp, and (3) a step function. The form of the functions $s(n)$s, $v(n)$s, and $x(n) = s(n) + v(n)$s are shown in Figure 8.3.

For simplicity, in our study, we shall use filters with odd number of ele-ments and, hence, $K = 2k + 1$. Under these circumstances the output of a **mean** filter at time n is given by

$$y(n) = \frac{1}{2k+1}\sum_{j=-k}^{k} x(n+j)$$

(8.59)

Since, in the above equation, for any n we average only $2k+1$ values of the signal sample from $-k$ to $+k$ with middle point n, the process is equivalent in multiplying the original signal with a window called the **filter window**. Because this window is moving with each new n, it is also known as the **moving filter window**. The signal samples which contribute to the output at time n are $x(n - k)$, $x(n - k + 1)$, ..., $x(n)$, ..., $x(n + k - 1)$, $x(n + k)$. For example, with filter length 5 and signal length 50, the output signal is given by

$$y(n) = \frac{1}{5}\sum_{j=-2}^{2} x(n+j), \quad n = 3, 4, \ldots, 48$$

(8.60)

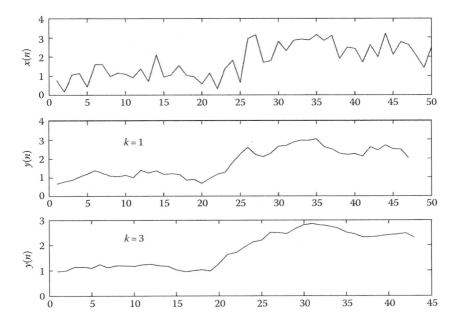

Figure 8.4

Figure 8.4 shows the noisy signal $x(n)$ and its filtered version for the mean filter of $k = 1$ and $k = 3$. The signal $s(n)$ is a step function with values 1 and 2.5. The random noise $v(n)$ is a WGN with mean zero. From the figure, we note that the output becomes **smoother** as the filter window increases, and the step change of the signal becomes a **ramp**. This indicates that the edges in images (abrupt change from white to black, for example) will be smoothed out, like looking at the image through a lens that is slightly out of focus.

Book MATLAB® function for one-dimensional mean filter:
[yo1] = ssp_1d_mean_filter(x,k)

```
function[yo1]=ssp_1d_mean_filter(x,k)
%filter length 2k+1;x=s+v=noisy signal
for n=1:length(x)-(2*k+1)
  for j=0:2*k
      y(j+1)=x(n+j);
  end;
  yo1(n)=sum(y)/(2*k+1);
end;
```

In a manner similar to the one-dimensional case, the center of a moving window is placed at every pixel and the mean operation uses the $(2k + 1)$

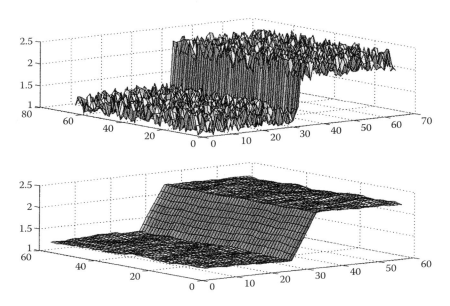

Figure 8.5

(2*k* + 1) pixel values inside the window and the result of the operation is the output at the window location.

Example 8.4

The three-dimensional signal used in this example is sum of a two-dimensional step function and white noise as shown in the upper part of Figure 8.5. The lower part of Figure 8.5 shows the output of a two-dimensional mean filter with sliding window dimensions 7×7.

The Book MATLAB functions and programs which produce Figure 8.5 are given below.

Book MATLAB function producing the noisy step signal:
[z] = ssp_2d_step_signal(N,M,i)

```
function[z]=ssp_2d_step_signal(N,M,i)
%N=points in y direction;M=points in x direction;
%i=controlling intensity of the noise;
%z=matrix presenting the step function and noise;
xy=i*rand(N,M);
xys=2.5*ones(N,M)+i*rand(N,M);
z=[xy xys];
```

Book MATLAB function that produces output from a mean filter:
[yo2] = ssp_2d_mean_filter(z,k)

```
function[yo2]=ssp_2d_mean_filter(z,k)
%z=2-dimensional signal with noise, matrix;
%2k+1=width and length of sliding window;
for i=1:length(z(1,:))-(2*k+1)
    for j=1:length(z(:,1))-(2*k+1)
        zw=z(i:i+2*k,j:j+2*k);
        yo2(i,j)=sum(sum(zw))/((2*k+1)^2);
  end;
end;
```

To produce Figure 8.5, we used the following MATLAB program:

```
N=64;M=32;i=0.5;
z=ssp_2d_step_signal(N,M,i);
k=3;
yo=ssp_2d_mean_filter(z,k);%yo=57×57 matrix;
[X1,Y1]=meshgrid(1:64);[X,Y]=meshgrid(1:57);
subplot(2,1,1);
surfl(X1,Y1,z);
subplot(2,1,2);
surfl(X,Y,yo);                                    ■
```

8.4   Median filter

When a single impulse exists in the signal, the mean filter spreads the impulse and reduces the amplitude. On the other hand, the median filter totally eliminates the impulse, provided that its width is less than $k + 1$ of a window length $N = 2k + 1$. Figure 8.6 shows the effect of mean and median filter on one-dimensional signal. Note the different effects on the input these two figures have.

Book MATLAB function for one-dimensional median filter:
[yo1] = ssp_1d_median(x,k)

```
function[yo1]=ssp_1d_median_filter(x,k)
%one-dimensional filter; 2k+1=width of
%sliding window;x=data vector to be filtered;
for n=1:length(x)-(2*k+1)
    w=x(n:n+2*k);
    yo1(n)=median(w);
end;
```

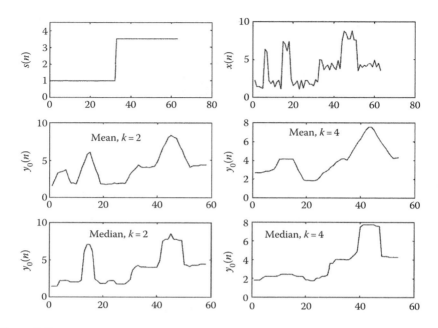

Figure 8.6

```
%the otput of the filter yo1 has length equal to
%length(x)-(2*k+1); median(.) is a MATLAB function;
```

The Book MATLAB filer for two-dimensional median filter is given below.

Book two-dimensional median MATLAB filter function:

```
function[yo2]=ssp_2d_median_filter(x,k)
%x=matrix (image);(2k+1)(2k+1)=number of
%matrix elements of sliding window;
for i=1:length(x(1,:))-(2*k+1)
  for j=1:length(x(:,1))-(2*k+1)
      xw=x(i:i+2*k,j:j+2*k);
      cvxw=xw(:);
      yo2(i,j)=median(cvxw);
  end;
end;
```

Figure 8.7 shows in the upper part the two-dimensional step signal with noise and a pulse. The middle figure shows the output of a median filter with *k* = 2. The bottom figure is the output of the median filter with *k* = 4. Note the complete disappearance of the pulse in the noisy signal.

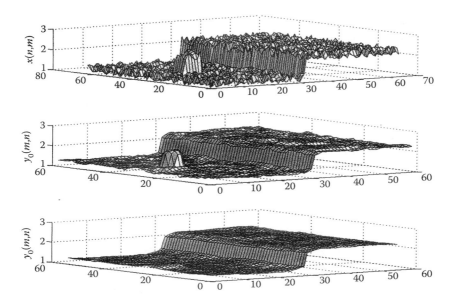

Figure 8.7

8.5 *Trimmed-type mean filter*

8.5.1 *(r − s)-Fold trimmed mean filters*

It can be easily shown by simulation that the mean filter is more efficient in deleting Gaussian noise than the median filter. However, the mean filter is less efficient in removing impulse noise. As long as the length of the impulse is less than $k + 1$, where $2k + 1$ is the width of the sliding window, the median filter completely eliminates the impulse noise. When both Gaussian and impulse noise are present, the **trimmed mean filter** becomes a compromise between mean and median filter.

One of the forms of the trimmed mean filter is the **($r−s$)-fold trimmed mean filter**, which is obtain by sorting the samples, omitting a total of $r+s$ samples: $x_{(1)}, x_{(2)}, ..., x_{(r)}$ and $x_{(N-s+1)}, x_{(N-s+2)}, ..., x_{(N)}$ and then average the remaining ones. The subscript with parenthesis indicates and ascending random values, e.g., $x_{(1)} \le x_{(2)} \le, ..., \le x_{(r)}$. Hence, we write

$$\text{Trimmed mean filter}\{x(1), x(2), ..., x(N); r, s\} = \frac{1}{N-r-s} \sum_{i=r+1}^{N-s} x_{(i)} \quad (8.61)$$

where $N = 2k + 1$ is the width of the sliding window.

The following Book MATLAB function executes the operation of a one-dimensional trimmed mean filter.

Book MATLAB for one-dimensional trimmed mean filter:
[yo1] = ssp_1d_trimmed_mean_filter(x,r,s,k)

```
function [yo1]=ssp_1d_trimmed_mean_filter(x,r,s,k)
%x=input noisy signal;r=integer;s=integer;
%r+s<N=2k+1=width of sliding window;
for n=1:length(x)-(2*k+1)
    for j=0:2*k
        y(j+1)=x(n+j);
    end;
    ys=sort(y);%sort(.)=MATLAB function;
    ystr=ys(r+1:2*k+1);
    yo1(n)=sum(ystr)/(2*k+1-r-s);
end;
```

Figure 8.8 compares the one-dimensional trimmed mean filter with median 1. We also observe the effect of changing the width of the sliding window.

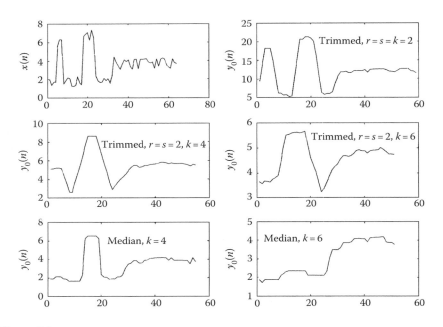

Figure 8.8

The following Book MATLAB function executes the two-dimensional trimmed mean filter.

Book MATLAB two-dimensional trimmed mean filter function:
[yo2] = ssp_2d_trimmed_mean_filter(x,r,s,k)

```
function[yo2]=ssp_2d_trimmed_mean_filter(x,r,s,k)
%x=input 2d signal (matrix);r=integer;s=integer;
%(2k+1)x(2k+1)=the number of elements inside the
%sliding window;r+s<2k+1;yo2=output matrix;
for i=1:length(x(1,:))-(2*k+1)
    for j=1:length(x(:,1))-(2*k+1)
        w=x(i:i+2*k,j:j+2*k);
        ws=sort(w(:));%sort(.) is a MATLAB function;
        wstr=ws(r+1:2*k+1-s);
        yo2(i,j)=sum(wstr)/(2*k+1-r-s);
    end;
end;
```

Figure 8.9 shows at the top the input signal to the filter. The middle figure shows the output of a mean filter with $k = 4$ and the bottom figure shows the output of a trimmed mean filter with $r = 2$, $s = 2$, and $k = 4$.

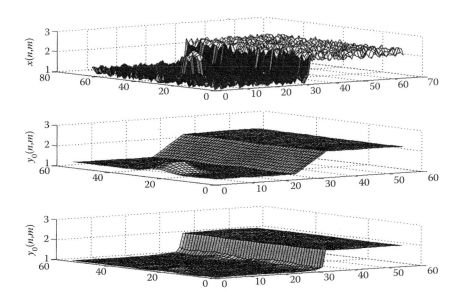

Figure 8.9

To plot, for example, the two-dimensional signal x (a 64 × 64 matrix), we write the program: `[X,Y] = meshgrid(1:64); surfl(X,Y,x); colormap gray;`.

8.5.2   (r,s)-Fold Winsorized mean filter

A modification of the (*r,s*)-fold trimmed mean filter is accomplished by substituting the values of the r smallest samples with the value $x_{(r+1)}$, and the values of the s largest samples are replaced by $x_{(N-s)}$, yielding

$$\text{Winsorized mean filter } \{x(1), \ldots, x(N); r,s\} = \frac{1}{N}\left(rx_{(r+1)} + \sum_{i=r+1}^{N-s} x_{(i)} + sx_{(N-s)}\right) \quad (8.62)$$

where $N = 2*k + 1$.

Book MATLAB function for one-dimensional (r,s)-fold Winsorized mean filter: [yo1] = ssp_1d_winsorized_mean_filter(x,r,s,k)

```
function[yo1]=ssp_1d_winsorized_mean_filter(x,r,s,k)
%x=input signal;r=integer;s=integer;r+s<N=2k+1=
%width of sliding window;
for n=1:length(x)-(2*k+1)
    for j=0:2*k
        y(j+1)=x(n+j);
    end;
  ys=sort(y);%sort(.) is a MATLAB function;
  ystr=ys(r+1:2*k+1-s);
  yo1(n)=(r*ys(r+1)+sum(ystr)+s*ys(2*k+1-s))/(2*k+1);
end;
```

Book MATLAB function for two-dimensional (r,s)-fold Winsorized mean filter: [yo2] = ssp_2d_winsorized_mean_filter(x,r,s,k)

```
function[yo2]=ssp_2d_winsorized_mean_filter(x,r,s,k)
%x=input 2d signal (image)=matrix;r=integer;s=integer;
%r+s<(2k+1)(2k+1)=number of elements inside the
%2d window;yo2=output matrix;
for i=1:length(x(1,:))-(2*k+1)
    for j=1:length(x(:,1))-(2*k+1)
        w=x(i:i+2*k,j:j+2*k);
        ws=sort(w(:));
        wstr=ws(r+1:2*k+1-s);
        yo2(i,j)=sum(wstr)/(2*k+1-s-r);
    end;
end;
```

8.5.3   Alpha-trimmed mean filter and alpha-Winsorized mean filter

In both cases, the trimmed elements are assumed to be equal in number, $r = s$. The trimmed number is often specified by a proportion denoted by $\alpha = j/N$, $0 \le j \le N/2$ is an integer. Hence, the number of samples trimmed at each side is αN. Since we have accepted N to be an odd number, we can select alpha to be even percentage such that αN is an even integer. The **α-trimmed filter** is given by

$$\text{Alpha-trimmed mean filter} \quad \{x(1), ..., x(N); a\} = \frac{1}{N - 2aN} \sum_{i=\alpha N+1}^{N-\alpha N} x_{(i)} \quad (8.63)$$

Book MATLAB function for one-dimensional alpha-trimmed mean filter:
[yo1] = ssp_1d_alpha_trimmed_mean_filter(x,a,k)

```
function[yo1]=ssp_1d_alpha_trimmed_mean_filter(x,a,k)
%x=data;a=proportion of trimmed elements;N=2k+1=
%number of elements of the sliding window;
N=2*k+1;
for n=1:length(x)-N
    for j=0:2*k
        y(j+1)=x(n+j);
    end;
  ys=sort(y);
  ystr=ys(a*N+1:N-a*N);
  yo1(n)=sum(ystr)/(N-2*a*N);
end;
```

Book MATLAB function for two-dimensional alpha-trimmed mean filter:
[yo2] = ssp_2d_alpha_trimmed_mean_filter(x,a,k)

```
function[yo2]=ssp_2d_alpha_trimmed_mean_filter(x,a,k)
%x=data;a=portion of trimmed elemnts;NxN=(2k+1)(2k+1)=
%number of elements inside the sliding window;yo2=
%filter output (matrix);a*N*N=must be an integer;
N=2*k+1;
for i=1:length(x(1,:))-N
    for j=1:length(x(:,1))-N
        w=x(i:i+2*k,j:j+2*k);
        ws=sort(w(:));
        wstr=ws(a*N*N+1:N*N-a*N*N);
        yo2(i,j)=sum(wstr)/(N*N-2*a*N*N);
    end;
end;
```

8.5.3.1   Alpha-trimmed Winsorized mean filter

The alpha-trimmed Winsorized mean filter is given by the relationship:

Alpha-trimmed wins mean filter

$$
\{x(1), x(2), ..., x(N); \alpha\} = \frac{1}{N}\left(\alpha N x_{(\alpha N+1)}\right) + \sum_{i=\alpha N+1}^{N-\alpha N} x_{(i)} + \alpha N x_{(N-\alpha N)}
\tag{8.64}
$$

Book MATLAB function for one-dimensional alpha-trimmed Winsorized mean filter: [yo1] = ssp_1d_alpha_trimmed_wins_mean_filter(x,a,k)

```
function[yo1]=ssp_1d_alpha_tr_wins_mean_filter(x,a,k)
%x=data;a=proportion of trimmed elements;N=k+1=number
%of elements inside the sliding window;a*N=must
%be an integer;
N=2*k+1;
for n=1:length(x)-N
    for j=0:2*k
        y(j+1)=x(n+j);
    end;
    ys=sort(y);
    ystr=ys(a*N*N+1:N-a*N);
    yo1(n)=(sum(ystr)+a*N*ys(a*N+1)+a*N*ys(N-a*N))/N;
end;
```

Book MATLAB function for two-dimensional alpha-trimmed Winsorized mean filter: [yo2] = ssp_2d_alpha_trimmed_wins_mean_filter(x,a,k)

```
function[yo2]=ssp_2d_alpha_tr_wins_mean_filter(x,a,k)
%x=data (matrix);a=portion of trimmed elements;
%NxN=(2k+1)(2k+1)=number of elements inside the
%sliding window;a*N*N=must be integer;
%yo2=output (matrix);
N=2*k+1;
for i=1:length(x(1,:))-N
    for j=1:length(x(:,1))-N
        w=x(i:i+2*k,j:j+2*k);
        ws=sort(w);
        wstr=ws(a*N*N+1:N*N-a*N*N);
        yo2(i,j)=sum(wstr)/(N*N-2*a*N*N);
    end;
end;
```

8.6   L-filters

The L-estimators are useful and are widely used because, by varying the associated constants, we obtain many useful estimators.

Definition 8.6   The L-estimators are of the form:

$$\text{L-estimator} \boxed{\hat{\theta} = \sum_{i=1}^{N} a_i x_{(i)}, \quad a_i s = \text{constants}} \tag{8.65}$$

The L-filters are also known as **order statistic filters**. They have been used by the statisticians for a long time since they have robust and often optimal properties for estimating population parameters of iid random variables. Furthermore, these filters are a compromise between nonlinear and linear operation since they include ordering and weighting. The L-filter is given by

$$\text{L-filter} \boxed{\{x(1), x(2), \ldots, x(N); \mathbf{a}\} = \sum_{i=1}^{N} a_i x_{(i)}, \quad \mathbf{a} = [a_1 \quad a_2 \quad \cdots \quad a_N]^T} \tag{8.66}$$

If in addition,

$$\sum_{i=1}^{N} a_i = 1 \tag{8.67}$$

The L-filter is known as the **smooth** L-filter. The great advantage of these filters is the ability to choose appropriate weighting factors to optimize the filtering in the mean square sense.

Book MATLAB function for one-dimensional L-filter:
[yo1] = ssp_1d_Lfilter(x,a,k)

```
function [yo1]=ssp_1d_Lfilter(x,a,k)
%x=input vector data;a=input weighting vector,
%length(a)=N;N=2k+1=length of sliding window;
N=2*k+1;
for n=1:length(x)-N
    for j=0:2*k
        y(j+1)=x(j+n);
    end;
    ys=sort(y);
    yo1(n)=sum(ys.*a);
end;
```

Book MATLAB function for two-dimensional L-filter:
[yo2] = ssp_2d_Lfilter(x,a,k)

```
function [yo2]=ssp_2d_Lfilter(x,a,k)
%x=input signal, matrix;a=input weighting matrix (NxN);
%NxN=(2k+1)(2k+1)=number of elements of the two-
%dimensional window;
N=2*k+1;
```

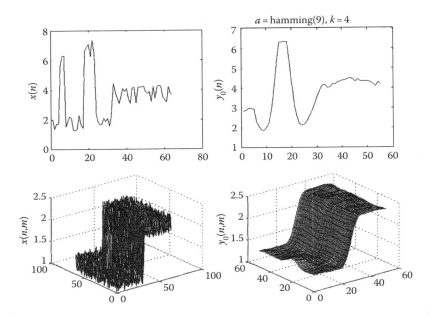

Figure 8.10

```
for i=1:length(x(1,:))-N
    for j=1:length(x(:,1))-N
        w=x(i:i+2*k,j:j+2*k);
        ws=sort(w(:));
        yo2(i,j)=sum(a(:)'.*ws');
    end;
end;
```

The two-dimensional output of Figure 8.10 was produced with a = hamming(13)*hamming(13)'; and $k = 4$.

8.7   Ranked-order statistic filter

The **rank-order** filters have been used by statisticians for a long time and are simple modifications of the median filter. Mathematically, we write

$$\text{Rank} - \text{order filter } \boxed{\{x(1),\ x(2),\ \ldots,\ x(N); r\} = x_{(r)}} \qquad (8.68)$$

which is the rth order statistic of the sample $\{x(1), x(2), \ldots, x(N)\}$. If $r = k + 1$, we obtain $x_{(k+1)}$ which acts as a median filter. If $r < k + 1$, a bias toward lower values is accomplished. If $r > k+1$ a bias toward higher values is accomplished.

Book MATLAB function for one-dimensional ranked-order filter:
[yo1] = ssp_1d_ranked_order_filter(x,r,k)

```
function[yo1]=ssp_1d_ranked_order_filter(x,r,k)
%x=input data;r=integer<=N=2k+1;N=length
%of sliding window;
N=2*k+1;
for n=1:length(x)-N
    for j=0:2*k
        y(j+1)=x(n+j);
    end;
    yst=sort(y);
    yo1(n)=yst(r);
end;
```

Book MATLAB function for two-dimensional ranked-order filter:
[yo2] = ssp_2d_ranked_order_filter(x,r,k)

```
function[yo2]=ssp_2d_ranked_order_filter(x,r,k)
%x=input data (matrix);r=integer<NxN;
%NxN=number of elements inside the sliding
%two-dimensional window; N=2k+1;
N=2*k+1;
for i=1:length(x(1,:))-N
    for j=1:length(x(:,1))-N
        w=x(i:i+2*k,j:j+2*k);
        ws=sort(w(:));
        yo2(i,j)=ws(r);
    end;
end;
```

Figure 8.11 shows the one-dimensional input signal $\{x(n)\}$ and the output for $r = 2$ and $r = 3$ both with the same integer $k = 6$. We note that a shift of the output takes place with different order statistic. Figure 8.12 shows the results of a two-dimensional filter with two different values of k.

8.8   Edge-enhancement filters

One type of edge-enhancement filters is the **comparison and selection** filter. Because the output of the median filter is smaller than the value of the output of the mean filter at the beginning of the signal edge, values of some smaller sample than the median is retained. Similarly, since the output of the median filter is larger than the output of the mean filter at the end of the signal edge, values of some larger samples than the median is retained. Formally, we write

Comparison selection filter $\{x(1), x(2), \ldots, x(N); j\} = \{x_{(k+1-j)},$ mean

$\{x(1), \ldots, x(N)\} \geq \text{med}\{x(1), \ldots, x(N)\}\}$ (8.69)

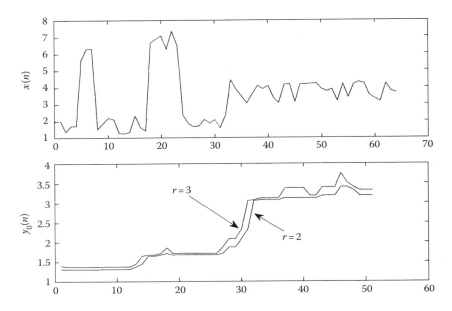

Figure 8.11

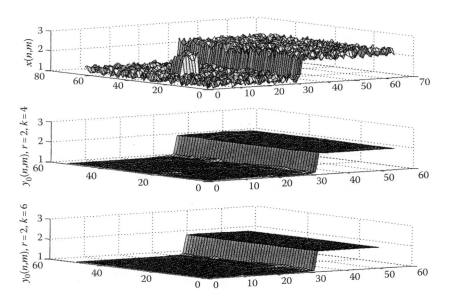

Figure 8.12

Book MATLAB function for one-dimensional comparison selection filter:
[yo1] = ssp_1d_comp_select_filter(x,j,k)

```
function[yo1]=ssp_1d_comp_select_filter(x,j,k)
%x=data;N=2k+1=length of sliding window;
%j=integer; 1<=j<k
N=2*k+1;
for n=1:length(x)-N
    for m=0:2*k
        y(m+1)=x(n+m);
    end;
    ys=sort(y);
    ym=mean(y);%mean()=MATLAB function;
    ymed=median(y);%median()=MATLAB function;
    if ym>=ymed
        yo1(n)=ys(k+1-j);
    else
        yo1(n)=ys(k+1+j);
    end;
end;
```

Book MATLAB function for two-dimensional comparison selection filter:
[yo2] = ssp_2d_comp_select_filter(x,j,k)

```
function[yo2]=ssp_2d_comp_select_filter(x,m,k)
%x=data (matrix); NxN=(2k+1)(2k+1)=dimensions
%of two-dimensional sliding window;m=integer,
%1<=m<k;
N=2*k+1;
for i=1:length(x(1,:))-N
    for j=1:length(x(:,1))-N
        w=x(i:i+2*k,j:j+2*k);
        ws=sort(w(:));
        wm=mean(ws);
        wmed=median(ws);
        if wm>=wmed
            yo2(i,j)=ws(k+1-m);
        else
            yo2(i,j)=ws(k+1+m);
        end;
    end;
end;
```

Figure 8.13 shows the effect of increasing sharpness of the edge of a signal with the increase of the parameter *j*.

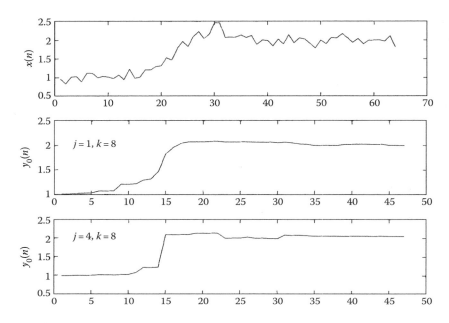

Figure 8.13

8.9   R-filters

R-filters are the result of R-estimators which are robust and are based on rank tests. The rank of an observation x_i is denoted by $R(x_i)$ and is given by

$$x_i = x_{(R(x_i))}, \quad i = 1, 2, \ldots, N \tag{8.70}$$

which means that the rank of x_i is in the ordered sequence. Next, we assign the weights

$$w_{jk} = \frac{d_{N-k+1}}{\sum_{i=1}^{N} i d_i} \tag{8.71}$$

to each of the $n(n+1)/2$ averages $(x_{(j)} + x_{(k)})/2$ for $j \leq k$. Then, the R-estimator is the median of the discrete distribution that assigns the probability w_{jk} to each average $(x_{(j)} + x_{(k)})/2$ for $j \leq k$. For example, if $d_1 = d_2 = \cdots = d_N = 1$ then the distribution assigns the weights $2/(N(N+1))$ to each of the averages $(x_{(j)} + x_{(k)})/2$ for $j \leq k$ and, thus,

$$\boxed{\begin{array}{l} \text{Hodges} - \text{Lehmann estimator } \{x_1, \ldots, x_N\} \triangleq \hat{\theta} \\ \quad = \text{med}\{(x_{(j)} + x_{(k)})/2 : 1 \leq j \leq k \leq N\} \end{array}} \tag{8.72}$$

known as the Hodges–Lehmann estimator.

Example 8.5

Let the input data be the vector x = [4   5   16   9   6].
Then, we sort the vector to obtain {x} = [4   5   6   9   16].
The output of the Hodges–Lehmann estimator is
med{(4 + 16)/2, (5 + 9)/2, (6 + 6)/2} = med{10, 7, 6} = 7.   ■

Book MATLAB one-dimensional Hodges–Lehmann filter:
[yo1] = ssp_1d_hodges_lehmann_filter(x,k)

```
function[yo1]=ssp_1d_hodges_lehmann_filter(x,k)
%x=data;N=2k+1=length of sliding window;
N=2*k+1;
for n=1:length(x)-N
    for m=0:2*k
        y(m+1)=x(n+m);
    end;
    ys=sort(y);
    for i=1:k+1
        aws(i)=(ys(i)+ys(N-i+1))/2;
    end;
    yo1(n)=median(aws);
end;
```

Book MATLAB two-dimensional Hodges–Lehmann filter:
[yo2] = ssp_2d_hodges_lehmann_filter(x,k)

```
function[yo2]=ssp_2d_hodges_lehmann_filter(x,k)
%x=data matrix;NxN=(2k+1)(2k+1)=dimensions of
%two dimensions sliding window;
N=2*k+1;
for n=1:length(x(1,:))-N
    for m=1:length(x(:,1))-N
        for m=1:length(x(:,1))-N
            w=x(n:n+2*k,m:m+2*k);
            ws=sort(w(:));
            for i=1:k+1
                aws(i)=(ws(i)+ws(N-i+1))/2;
            end;
            yo2(n,m)=median(aws);
        end;
    end;
end;
```

Figure 8.14 shows the result of the Hodges–Lehmann on a noisy two-dimensional signal with *k* = 2 and *k* = 4.

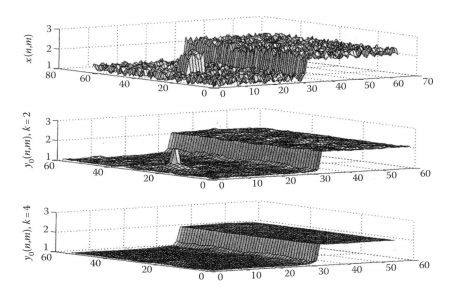

Figure 8.14

Problems, solutions, suggestions, and hints

8.2.1 Verify Equation 8.11.

Solution

$$\operatorname{var}\{\hat{c}\} = \operatorname{var}\left\{\frac{1}{N}\sum_{n=0}^{N-1}x(n)\right\} = E\{[\hat{c} - E\{\hat{c}\}]^2\} = E\left\{\left[\frac{1}{N}\sum_{n=0}^{N-1}x(n) - c\right]^2\right\}$$

$$= E\left\{\left(\frac{1}{N}\sum_{n=0}^{N-1}x(n)\right)^2 + c^2 - 2c\frac{1}{N}\sum_{n=0}^{N-1}x(n)\right\}$$

$$= E\left\{\left(\frac{1}{N}\sum_{n=0}^{N-1}x(n)\right)^2\right\} + c^2 - 2cE\left\{\frac{1}{N}\sum_{n=0}^{N-1}x(n)\right\}$$

$$= E\left\{\left(\frac{1}{N}\sum_{n=0}^{N-1}x(n)\right)^2\right\} - c^2 \left(\text{since } E\left\{\frac{1}{N}\sum_{n=0}^{N-1}x(n)\right\} = \frac{1}{N}\sum_{n=0}^{N-1}E\{x(n)\} = \frac{Nc}{N} = c\right)$$

$$= E\left\{\frac{1}{N^2}\sum_{n=0}^{N-1}x^2(n) + \frac{1}{N}\sum_{n\neq m}x(n)x(m)\right\} - c^2 = \frac{1}{N^2}\sum_{n=0}^{N-1}E\{x^{2(n)}\} + \frac{1}{N}\sum_{n=0}^{N-1}c^2 - c^2$$

$$= \frac{1}{N^2}\sum_{n=0}^{N-1}\operatorname{var}\{x(n)\} + \frac{Nc}{N} - c^2 = \frac{N\sigma^2}{N^2} = \frac{\sigma^2}{N}, \qquad x(n),\ x(m)\text{ are independent}$$

8.2.2 Let $v(n)$ be a WGN with zero mean and equal variance. If the observation is $x(n) = c + v(n)$ for $n = 0, 1, \ldots, N - 1$, find the estimate of the parameter c.

Solution

If we set

$$\hat{c} = \frac{1}{N}\sum_{n=0}^{N-1} x(n) = g(x(0), x(1), \ldots, x(N-1)) \tag{1}$$

then

$$E\{\hat{c}\} = \frac{1}{N}\sum_{n=0}^{N-1} E\{x(n)\} = \frac{1}{N}\sum_{n=0}^{N-1} E\{c + v(n)\} = \frac{1}{N}\sum_{n=0}^{N-1} (c + E\{v(n)\}) = \frac{Nc}{N} = c$$

for all c. This implies that the estimator (1) is unbiased.

8.2.3 Show that $\hat{c} = (1/4N)\sum_{n=0}^{N-1} x(n)$, with $x(n) = c + v(n)$ (see also Problem 8.2.2) is a biased estimator.

Solution

$$E\{\hat{c}\} = \frac{1}{4N}\sum E\{x(n)\} = \frac{1}{4}c$$

which indicates that if $c = 0$ then $E\{\hat{c}\} = c$ and if $c \neq 0$ then $E\{\hat{c}\} \neq c$ and, thus, $\hat{c}$ is an unbiased estimator.

8.2.4 Verify Equation 8.27.

Solution

Taking the partial derivative of Equation 8.26 with respect to a_m, we find

$$\partial \mathrm{mse}(\hat{\theta})/\partial a_m = -2E\{\theta x(m)\} + \frac{\partial}{\partial a_m}\left[\sum_{m=0}^{N-1}\sum_{n=0}^{N-1} a_m a_m E\{x(m)x(n)\}\right]$$

$$= -2r_{\theta x}(m) + 2\sum_{n=0}^{N-1} a_n r_{mn} = 0 \tag{1}$$

since $a_m a_n$ appears twice in the double summation expansion. Therefore, (1) is the mth equation of Equation 8.27.

8.2.5 Let $x(n) = s(n) + v(n)$ for $n = 0, 1, \ldots, N - 1$, where $v(n)$s are iid with zero mean and variance σ_n^2 for $n = 0, 1, \ldots, N - 1$. Find the as in Equation 8.27.

Solution

$$
\begin{aligned}
r_{mn} &= E\{x(m)x(n)\} = E\{(s(m)+v(m))(s(n)+v(n))\} \\
&= E\{(s(m)s(n)\} + E\{v(m)s(n)\} + E\{s(m)v(n)\} + E\{v(m)v(n)\}
\end{aligned}
\tag{1}
$$

If

$$
m \neq n \; r_{mn} = \sigma_s^2
\tag{2}
$$

and if

$$
m = n \; r_{mn} = \sigma_s^2 + \sigma_v^2
\tag{3}
$$

Also

$$
r_{sx}(m) = E\{s(m)s(m) + s(m)v(m) = \sigma_s^2 + E\{s(m)v(m)\} = \sigma_s^2
\tag{4}
$$

Hence $Ra = r_{\theta x}$ where

$$
\mathbf{R} =
\begin{bmatrix}
\sigma_s^2 & \sigma_s^2 & \cdots & \sigma_s^2 \\
\sigma_s^2 & \sigma_s^2 & \cdots & \sigma_s^2 \\
\vdots & \vdots & \vdots & \vdots \\
\sigma_s^2 & \sigma_s^2 & \cdots & \sigma_s^2
\end{bmatrix}
+
\begin{bmatrix}
\sigma_0^2 & 0 & \cdots & 0 \\
0 & \sigma_1^2 & \cdots & 0 \\
\vdots & \vdots & \vdots & \vdots \\
0 & 0 & \cdots & \sigma_{N-1}^2
\end{bmatrix},
\quad
\mathbf{a} \triangleq
\begin{bmatrix}
a_0 \\
\vdots \\
a_{N-1}
\end{bmatrix}
= \mathbf{R}^{-1}
\begin{bmatrix}
\sigma_s^2 \\
\vdots \\
\sigma_s^2
\end{bmatrix}
$$

8.2.6 Let the rvs $x(0), x(1), \ldots, x(N - 1)$ denote a random sample, are iid and the pdf is

$$
p(x(m)) =
\begin{cases}
\theta^{x(m)}(1-\theta)^{1-x(m)}, & x(m) = 0,1, \; 0 \leq \theta \leq 1 \\
0, & \text{elsewhere}
\end{cases}
$$

Find the maximum likelihood function of θ.

Solution

The likelihood function is

$$
L(\theta; x(0), \ldots, x(N-1)) = \theta^{\Sigma x(m)}(1-\theta)^{N-\Sigma x(m)}
\tag{1}
$$

where $x(m)$ are 0 and 1 for $m = 0, 1, \ldots, N - 1$.

The natural logarithm of (1) is

$$\ln L(\theta) = \sum_{m=0}^{N-1} x(m)\ln\theta + N - \sum_{m=0}^{N-1} x(m) + \ln(1-\theta)$$

Hence,

$$\partial\ln L(\theta)/\partial\theta = \sum_{m}^{N-1} x(m)(1/\theta) - \left(N - \sum_{m}^{N-1} x(m)\right)/(1-\theta) = 0 \tag{2}$$

provided that θ is not 0 or 1. Solving (2) for θ, we obtain the maximum likelihood estimator:

$$\hat{\theta} = \sum_{m=0}^{N-1} x(m)/N$$

If we set $N = 3$, $x(0) = 0$, $x(1) = 1$, $x(2) = 1$ then $L(\theta) = \theta^2$ $(1-\theta)^{3-2} = \theta^2(1-\theta)$ and $\hat{\theta} = 2/3$.

8.2.7 A set of data is given by $x(n) = c + v(n)$ for $n = 0, 1, ..., N-1$, where $v(n)$ is a WGN with zero mean and known variance σ^2. Show that the estimator $\hat{c}$ is an efficient estimator.

Solution

The pdf is

$$p(x;c) = p(x(0);c)p(x(1);c)\cdots p(x(N-1);c) = \left[\frac{1}{(2\pi\sigma^2)^{N/2}}\right]\exp\left[-\frac{1}{2\sigma^2}\sum_{n=0}^{N-1}(x(n)-c)^2\right] \tag{1}$$

The derivative of the logarithm of the likelihood function is

$$\frac{\partial\ln p(x;c)}{\partial c} = \frac{1}{\sigma^2}\sum_{n=0}^{N-1}(x(n)-c) = 0 \tag{2}$$

which yields the estimator:

$$\frac{1}{\sigma^2}\sum_{n=0}^{N-1}x(n) = \frac{1}{\sigma^2}N\hat{c} \quad \text{or} \quad \hat{c} = \frac{1}{N}\sum_{n=0}^{N-1}x(n) \tag{3}$$

From (2), we also obtain

$$\frac{\partial^2 \ln p(\mathbf{x};c)}{\partial c^2} = \frac{N}{\sigma^2}$$

which shows that the second derivative is a constant. From (2), we also obtain the mean value to be

$$E\{\hat{c}\} = \frac{1}{N}\sum_{n=0}^{N-1} E\{x(n)\} = (1/N)Nc = c$$

which shows that the estimator is *unbiased*. Since the Cramer–Rao formula gives the relation $1/E\{-\partial^2 \ln p(x;c)/\partial c^2\} = \sigma^2/N$ indicates that the variance of the estimator is var $(\hat{c}) = \sigma^2/N$ which attains the lowest value, known as the CRLB, as $N \to \infty$. Hence, the estimator is *asymptotically efficient*.

8.2.8   Let us arrange a set of rvs $\{x(0)\ x(1)\ x(2)\ x(3)\ x(4)\}$ in ascending order of their values $\{x_{(0)}\ x_{(1)}\ x_{(2)}\ x_{(3)}\ x_{(4)}\}$. The population of this set has a Laplacian pdf. Find the estimator.

Solution

Referring to Figure 8.15 the sum $|x_{(0)} - \theta| + |x_{(4)} - \theta|$ becomes minimum in the range $x_{(1)} \le \theta \le x_{(4)}$ and remains constant in this range. The sum $|x_{(1)} - \theta| + |x_{(3)} - \theta|$ becomes minimum in the range $x_{(2)} \le \theta \le x_{(4)}$ and stays constant. The last factor of the summation $\sum_{n=0}^{4-1}|x(n) - \theta|$ is $|x_{(2)} - \theta|$ and this factor becomes minimum if $\theta = x_{(2)}$, where $x_{(2)}$ is the median of the set.

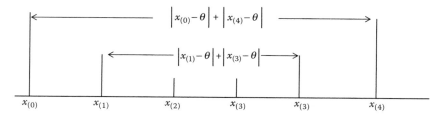

Figure 8.15

8.3.1   Show that the mean filter is linear.

Solution

The linearity is defined by $f(ax + by) = af(x) + bf(y)$, where a and b are real or complex constants. From the definition, we obtain

$$y(n) = \frac{1}{2k+1} \sum_{j=-k}^{k} x(n+j) = \frac{1}{2k+1} \sum_{j=-k}^{k} [s(n+j) + v(n+j)]$$

$$= \frac{1}{2k+1} \sum_{j=-k}^{k} s(n+j) + \frac{1}{2k+1} \sum_{j=-k}^{k} v(n+j)$$

$$= (\text{filtered signal}) + (\text{filtered noise})$$

8.4.1   Find the output of a median filter if the input sequence is $x = \{1\,9\,4\,3\,8\,7\,7\}$, and $k = 1$.

Solution

$N = 2*1 + 1 = 3$, yo1 $= \{\text{med } \{1\,9\,4\} \text{ med}\{9\,4\,3\} \text{ med}\{4\,3\,8\}$
med$\{3\,8\,7\}\} = \{4\,4\,4\,7\}$.

8.5.1   Find the output of a trimmed mean filter if the input is $x = \{2\,6\,1\,5\,7\,9\,8\}$ and $r = 1$, $s = 1$, and $k = 2$.

Solution

$$xs = \{1\,2\,5\,6\,8\,9\}, \quad \text{yo1} = \left\{\frac{2+5+6}{5-1-1}, \frac{5+6+7}{3}, \frac{6+7+8}{3}\right\} = \{4.3\,6\,7\}$$

8.5.2   Find the output of a (r,s)-fold Winsorized mean filter if the input is $x = \{2\,6\,1\,5\,7\,9\,8\}$ and $r = 1$, $s = 1$, and $k = 2$.

Solution

$xs = \{1\,2\,5\,6\,7\,8\,9\}$, yo1(1) $= (1 \times 2 + (2+5+6) + 1 \times 6)/5 = 4.2$,
yo1(2) $= 6$.

8.5.3   Find the output of an alpha-trimmed mean filter if the input is $x = \{2\,6\,1\,5\,7\,9\,8\}$ and $\alpha = 0.2$, $k = 2$.

Solution

$xs = \{1\,2\,5\,6\,7\,8\,9\}$, $N = 2 \times 2 + 1 = 5$, $\alpha N = 1$. Hence, yo1(1)
$= (2 + 5 + 6)/(5 - 2 \times 0.2 \times 5) = 4.33$, yo1(2) $= (5 + 6 + 7)/3$
$= 6$, yo1(3) $= 7$.

8.6.1 Find the output of an L-filter if $x = \{2\ 1\ 6\ 4\ 5\ 9\ 8\ 7\ 3\}$, $a = [0.4\ 0.2\ 0.4]$ and $k = 1$.

Solution

yo1(1) = 1 × 0.4 + 2 × 0.2 + 6 × 0.4 = 3.2, yo1(2) = 1 × 0.4 + 4 × 0.2 + 6 × 0.4 = 3.6, yo1(3) = 5.0, yo1(4) = 6.2, yo1(5) = 7.2, and yo1(6) = 8.0.

8.7.1 Find the output of a ranked-order filter with $r = 2$ and $k = 2$. The input data are $x = \{5\ 1\ 3\ 9\ 8\ 7\ 6\ 2\ 4\}$.

Solution

$N = 5$, yo1s(1) = $\{1\ 3\ 5\ 8\ 9\}$, yo1(1) = $x_{(2)} = 3$, yo1(2) = $\{1\ 3\ 6\ 7\ 8\ 9\}$, yo1(2) = $x_{(2)} = 3$, yo1s(3) = $\{2\ 6\ 7\ 8\ 9\}$, yo1(3) = $x_{(2)} = 6$, etc.

Bibliography

Astola, J. and P. Kuosmanen, *Fundamentals of Nonlinear Digital Filtering*, CRC Press, Boca Raton, FL, 1997.

Farhang-Boroujeny, B., *Adaptive Filters: Theory and Applications*, John Wiley & Sons, New York, NY, 1999.

Haykin, S., *Adaptive Filter Theory*, Prentice Hall, Upper Saddle River, NJ, 2001.

Hays, M. H., *Statistical Digital Signal Processing and Modeling*, John Wiley & Sons, New York, NY, 1996.

Hogg, R. V. and A. T. Craig, *Introduction to Mathematical Statistics*, 4th edn., Macmillan, New York, NY, 1978.

Kay, S. M., *Modern Spectrum Estimation: Theory and Applications*, Prentice Hall, Englewood Cliffs, NJ, 1988.

Kay, S. M., *Statistical Signal Processing*, Prentice Hall, Upper Saddle River, NJ, 1993.

Manolakis, D. G., V. K. Ingle, and S. T. Kogon, *Statistical and Adaptive Signal Processing*, McGraw-Hill, New York, NY, 2000.

Marple, S. L. Jr., *Digital Spectral Analysis with Applications*, Prentice Hall, Englewood Cliffs, NJ, 1987.

Pitas, I. and A. N. Venetsanopoulos, *Nonlinear Digital Filters: Principles and Applications*, Kluwer Academic Publishers, Boston, MA, 1990.

Poularikas, A. D. and Z. M. Ramadan, *Adaptive Filtering Primer with MATLAB*, CRC/Taylor & Francis, Boca Raton, FL, 2006.

Poularikas, A. D., *Signals and Systems Primer with MATLAB*, CRC/Taylor & Francis, Boca Raton, FL, 2007.

Shiavi, R., *Introduction to Applied Statistical Signal Analysis*, Irwin, Boston, MA, 1991.

Stoica, P. and R. Moses, *Introduction to Spectral Analysis*, Prentice Hall, Upper Saddle River, NJ, 1997.

Appendix A

Suggestions and explanations for MATLAB® use

It is suggested that the reader, who does not have much experience with MATLAB and before start using the text, goes over this appendix and tries to execute the presented material in MATLAB.

Creating a directory

It was found by the author that it is less confusing if for a particular project we create our own directory where our own developed MATLAB m-files are stored. However, any time we need anyone of these files, we must include the directory in the MATLAB path. Let us assume that we have the following directory path: c:\ap\sig-syt\ssmatlab. The following two approaches can be used

```
        >>cd 'c:\ap\sig-syst\ssmatlab'
or
        >>path(path, 'c:\ap\sig-syst\ssmatlab')%remember
%to introduce the path any time you start new
%MATLAB operations; the symbol is necessary
        % for the MATLAB to ignore the explanations;
```

The MATLAB files are included in "ssmatlab" directory.

Help

In case we know the name of a MATLAB function and we would like to know how to use it, we write the following command in the command window:

```
        >>help sin
or
        >>help exp
```
etc.

For the case we want to look for a keyword, we write

```
        >>look for filter
```

Save and load

When we are in the command window and we have created many variables and, for example, we would like to save two of them in a particular directory and in a particular file, we proceed as follows:

```
>>cd 'c:\ap\matlabdata'
>>save data1 x dt %it saves in the matlabdata
  %directory the
%file data1 having the two variables x and dt;
```

Let us assume now that we want to bring these two variables in the working space to use them. We first change directory, as we did above, and then we write in the command window:

```
>> load data1
```

Then, the two variables will appear in the working space ready to be used.

MATLAB as calculator

```
>>pi^pi-10;
>>cos(pi/4);
>>ans*ans %the result will be  (√2̄/2)×(√2̄/2)=1/2
  because the
          %first output is eliminated, only the
          %last output is
          %kept in the form of ans;
```

Variable names

```
>>x=[1 2 3 4 5];
>>dt=0.1;
>>cos(pi*dt);%since no assignment takes place there
%is no variable;
```

Complex numbers

```
>>z=3+j*4;%note the multiplication sign;
>>zs=z*z;%or z^2 will give you the same results;
>>rz=real(z);iz=imag(z):%will give rz=3, and iz=4;
>>az=angle(z); abz=abs(z);%will give az=0.9273
%rad, and abz=5;
>>x=exp(-z)+4;%x=3.9675+j0.0377;
```

Array indexing

```
>>x=2:1:6;%x is an array of the numbers {2, 3, 4,
%5, 6};
>>y=2:-1:-2:%y is an array of the numbers {2, 1, 0,
%-1, -2};
>>z=[1 3 y];%z is an array of the numbers {1, 3,
%2, 1, 0, -1, -%2};
%note the required space between array
%numbers;
>>xt2=2*x;%xt2 is an array of numbers of x each
%one multiplied by 2;
>>xty=x.*y;%xty is an array of numbers which are
%the result of
            %multiplication of corresponding
            %elements, that is
            %{4, 3, 0, -5, -12};
```

Extracting and inserting numbers in arrays

```
>>x=2:1:6;
>>y=[x zeros(1,3)];%y is an array of the numbers
                    %{2, 3, 4, 5, 6, 0, 0, 0};
>>z=y(1,3:7);%1 stands for row 1 which y is and
            %3:7 instructs to
            %keep columns
            %3 through 7 the result is the array
            %{4, 5, 6, 0, 0};
lx=length(x);%lx is the number equal to the
            %number of columns of
            %the row vector x, that is lx=5;
x(1,2:4)=4.5*(1:3);%this assignment substitutes
        %the elements of x
        %at column
        %positions 2,3 and 4 with the numbers 4.5*
        %[1 2 3]=4.5, 9,
        %and 13.5, note the columns of 2:4 and 1:3
        %are the same;
x(1,2:2:length(x))=pi;% substitutes the columns 2 and
    %4 of x with
    %the value of pi, hence the array is {2, 3.1416,
    %4,3.1416 6}
```

Vectorization

```
>>n=0:0.2:1;
>>s=sin(0.2*pi*n);% the result of these two
      %commands gives the
      %signal s (sine function) at times (values
      %of n)0, 0.2,
      %0.4, 0.6, 0.4, 1;
```

This approach is preferable since MATLAB executes faster than the vectorization approach rather than the loop approach, which is

```
>>s=[];% initializes all values of vector s to
%zero;
>>for n=0:5% note that the index must be integer;
>>s(n+1)=sin(0.2*pi*n*0.2);% since we want values
            %of s every 0.2
            %seconds we must multiply n by 0.2; note
            %also that
            %for n=0 the variable becomes s(1) and
            %this
            %because the array in MATLAB always
            %starts
            %counting columns from 1;
>>end;
```

The results are identical with the previous one.

Matrices

If **a** and **b** are matrices such that **a** is a 2×3 and **b** is 3×3 then $c = a*b$ is a 2×3 matrix.

```
>>a=[1 2; 4 6]; %a is a 2x2 matrix
```
$\begin{bmatrix} 1 & 2 \\ 4 & 6 \end{bmatrix}$;

```
>>b=a';%b is a transposed 2x2 matrix of
%a and is
```
$\begin{bmatrix} 1 & 4 \\ 2 & 6 \end{bmatrix}$;

```
>>da=det(a);%da is a number equal to the
  %determinant of a, da=-2;
>>c=a(:);%c is a vector which is made up of the
%columns of a,
          %c=[1 4 2 6];
>>ia=inv(a); %ia is a matrix which is the inverse
%of a;
```

```
>>sa1=sum(a,1);%sa1 is a row vector made up of
%the sum of the
             %rows,sa1=[5 8];
>>sa2=sum(a,2);%sa2 is a column vector made up
    %by the sum of the
    %columns,sa2=[3 10]';
```

Produce a periodic function

```
>>x=[1 2 3 4];
>>xm=x'*ones(1,5);%xm is 4x5 matrix and each of
%its column is x';
>>xp=xm(:)';% xp is a row vector, xp=[x x x x x];
```

Script files

Script files are m-files that when introduce their names in the command window we receive the results. We must, however, have the directory that includes the file in the MATLAB search directories. You can modify the file in any desired way and get new results. Suppose that any time we ask for the file pexp.m the magnitude and angle of the exponential function $e^{j\omega}$ are plotted. To accomplice this, we first go to the command window and open a new m-file. At the window, we type the file as shown below. As soon as we finished typing, we click on "Save as" and save the file in, say:c:\a\pssmatlab. If we want to see the results, at the command window we just write pexp and hit the enter key.

Script file pexp.m

```
>>w=0:pi/500:pi-pi/500;%they are 500 at pi/500
%appart;
>>x=exp(j*w);ax=abs(x);anx=angle(x);
>>subplot(2,1,1);plot(w,ax,'k')%'k' means plot
%line in black;
>>xlabel('omega rad/s');ylabel
('Magnitude');
>>subplot(2,1,2);plot(w,anx,'k');
>>xlabel('omega rad/s');ylabel('Angle');
```

If we have the function $(2e^{j\omega}/(e^{j\omega}-0.5))$ and want to plot the results as above, we substitute in the script file the function x with the function: $x = 2*exp(j*w)/$ $(exp(j*w) - 0.5)$.

In the above MATLAB expression note the dot before the slash. This instructs MATLAB to operate at each value of w separate and, thus, give results at each frequency point.

Functions

We will present here an example of how to write functions. The reader should also study the functions which are presented through out the book. In Fourier series, for example, we have to plot functions of the form:

$$s(t) = \sum_{n=0}^{N} A_n \cos n\omega_0 t$$

and we want to plot this sum of cosines each one having different amplitude and frequency. Let A = [1 0.6 0.4 0.1], ω_0 = 2 and $0 \leq t \leq 4$. We approach this solution by vectorizing the summation. The MATLAB function is of the form:

```
function [s]=sumofcos(A,N,w0,rangeoft)
n=0:N-1;
s=A*cos(w0*n'*rangeoft)
```
%when we want to use this function at the command window
%to find s we write for example:
>>A=[1 0.6 0.4 0.1];N=4;w0=2;rangeoft=0:0.05:6;
>>[s] = sumofcos(A,N,w0,rangeoft);
at the enter key click the vector s is one of the
%variables in the command window and it can be plotted at
the wishes of the reader; we must secure that the directory
in which sumofcos function exists is in the MATLAB path;
after you type the function in the editing window you
"save as" in the directory, for example, c:\ap\ssmatlab and
filename: sumofcos.m

It is recommended that the reader set small numbers for N (N = 4) and range of t (0:0.2:1) and produce first the matrix $\cos(w_0*n'*t)$ and, then, see the result $A*\cos(w_0*n'*t)$.

Complex expressions

We can produce results by writing, for example,
>>x=[1 3 1 5 3 4 5 8];
>>plot(abs(fft(x,256)),'r');%will plot in red color the
                            %spectrum of the
                            %vector x of 256 points;

Axes

>>axis([xmin xmax ymin ymax]);%sets the max and min values
%of the axes;
>>grid on;%turns on grid lines in the graph;

Two-dimensional (2D) graphics

To plot a sine and a cosine signal
```
>>x=linspace(0,2*pi,40);%produces 40 equal spaced points
                        %between 0 and
                        %2π;
>>y=sin(x);plot(x,y,'r');%will plot the sine signal with
                        %color red;
>>y1=cos(x);plot(x,y1,'g');%will plot the cosine signal
                        %with color
                        %green;
```

For other color lines: 'y'=yellow,'c'=cyan,'b'=blue,'w'=white,'k'= black

Type of lines :'g:'=green dotted line,'r--'=red dashed line, 'k--x'=black
dashed line with x's,'k-.'=black dash-dot line, '+'=plus sign,'ko'=black circles

Add Greek letters: \omega=will produce Greek lower case omega, \Omega= will produce capital case Greek omega. The same is true for the rest of the Greek letters. For example if we want to write the frequency in a figure under the x-axis, in the command window we wite:>>xlabel('\omega rad/s');. For an omega with a subscript 01 we write:>>xlabel('\omega_{01} rad/s');

Add grid lines: >>grid on;%this is done after the command plot;

Adjusting axes: >>axis square;%sets the current plot to be %square rather than the default rectangle;
>>axis off;%turn off all axis labeling,
%grid,and tick marks;leave the title and
%any labels placed by the 'text'and
%'gtext' commands;
>>axis on;%turn on axis labaling,tick marks
%and grid;
>>axis([xmin xmax ymin ymax]);set the
%maximum and minimum values of the axes
%using values given in the row vector;

Subplots (Example): >>n=0:100;x=sin(n*pi*n);y=cos(n*pi*n);z=x.*y;...
>>w=x+y;
>>subplot(2,2,1);plot(n,x);subplot(2,2,2);...
>>plot(n,y);
>>subplot(2,2,3);plot(n,z);subplot(2,2,4);...
>>plot(n,w);

Log plotting: >>semilogx(x);%will plot the vector x in log
%scale in x-axis
%and linear scale in y-axis;
>>semilogy(x);%will plot the vector x in log
%scale in y-direction and linear scale in
%the x-axis;
>>loglog(x);%will plot the vector x in log
%scale both axes;

Histogram: >>x=randn(1,1000);hist(x,40);colormap([0 0 0]);%will
%plot a Gausian histogram of 40 bars white;if
%instead we entered the vector [1 1 1] the
%bares would be black;the vector [1 0 0] will
%give red and the vector [0.5 0.5 0.5] will
%give gray;
>>x=-3.0:0.05:3;y=exp(-x.*x);bar(x,y);...
>>colormap([.5 .5 .5]); %will produce bar-
%figure of the bell curve with gray color;
>>**sairs**(x,y,'k');%will produce a stair-like
%black curve;

Add words: >>gtext('the word');

After the return, the figure will appear and a crosshair. Move the center at
the point in the figure where the word must start and click.

Add legend: >>plot(x1,y1,'+',x2,y2,'*');%there will be two
%curves in the graph;
>>legend('Function 1','Function 2');

The following rectangle will appear in the figure:

+ Function1
* Function2

Three-dimensional (3D) plots

Mesh-type figures
If, for example, we desire to plot the function $f(x) = e^{-(x^2 + y^2)}$ in the ranges
$2 \le x \le 2, 2 \le y \le 2$, we proceed as follows:

```
>>x=-2:0.1:2;y=-2:0.1:2;[X,Y]=meshgrid(x,y);
>>f=exp(-(X.*X+Y.*Y));mesh(X,Y,f);colomap([0 0 0]);
```

The above commands will produce a mesh-type 3D figure with black
lines.

A.1 General purpose commands

Managing commands and function

help   On line help for MATLAB functions and m-files, e.g., >> help plot
path   Shows the path to MATLAB directories which are available at the command window

Managing variables and the workplace

clear   Removes all the variables and items in the memory. Let assume that the memory contains the variables x, y, z then >>clear x z; only y will remain in the memory
length   A number that gives the length of a vector. >>x = [1 3 2 5]; then >>length(x); will give the number 4. If we write >>y = length(x); then the variable y is equal to 4.
size   Array dimensions. >>x = [1 3 2 5]; then size(x) will give the numbers 1 4 which means 1 row and 4 columns. Let write >>x = [1 2; 3 5; 6 4]; then size(x) will give the numbers 3 2, which means that x is a matrix of 3 × 4 dimensions
who   Produces a list of the variables in the memory
format   This command is used as follows for display: >>format short,pi; will produce the number 1.1416, >>format long,pi; will produce the number 3.14159265358979, >>format long,single(pi); will produce the number 3.1415927

A.2 Operators and special characters

Operators and special characters

+   Plus
−   Minus
*   Number and matrix multiplications
.*   Array multiplication. >>x = [1 2 3]; y = [2 3 4]; z = $x.*y$; hence z = [2 6 12]
.^   Array power. >>x = [2 3 4]; $y = x.\text{\textasciicircum}3$; hence y = [8 27 64]. >>x = [2 4; 1 5]; $y = x.\text{\textasciicircum}2$; hence y = [4 16; 1 25]
/   Right division
./   Array division. >>x = [2 4 6]; y = [4 4 12]; z = $x./y$; hence z = [0.5 1 0.5]
:   Colon. >>x = [1 3 6 2 7 8]; y = $x(1,3:6)$; hence y = [6 2 7 8]
.   Decimal point
...   Continuation. >>x = [1 4 6 7 8 9 ... >>2 5 8 1]; the vector x is interpreted by MATLAB as a row vector having 10 elements
%   Comments. >>x = [1 4 2 6]; %this is a vector. MATLAB ignores "this is a vector"

'   Transpose of a matrix or vector. $>>x = [2\ 6\ 3]; y = x';$ will have $y = \begin{bmatrix} 2 \\ 6 \\ 3 \end{bmatrix}$

&   Logical AND
| Logical OR
~   Logical NOT
xor   Logical exclusive (XOR)

Control flow

for   Repeat statements a specific number of times.
       >>for n=0:3;
       >> $x(n+1)$=sin(n*pi*0.1); %observe the $n+1$, if the +1 was not there $x(0)$
       >>end; %was not defined by MATLAB

       Then x=[0 0.3090 0.5878 0.8090]

       >>for n=0:2
       >> for m=0:1
       >> $x(n+1,m+1)$=n^2+m^2;
       >> end;
       >>end;

       Then $x = \begin{bmatrix} 0 & 1 \\ 1 & 2 \\ 4 & 5 \end{bmatrix}$

while Repeat statements an indefinite times of times.
       >>a=1; num=0;
       >>while (1+a)<=2 & (1+a)>=1.0001
       >>a=s/2;
       >>num=num+1;
       >>end;
       We obtain a=0.0001, and num=14

if   Conditionally execute statements.
       if expression
               commands evaluated if true
       else
               commands evaluated if false
       end

If there are more than one alternative, the if-else-end statement takes the form

elseif
       if expression 1
               commands evaluated if expression 1 is true

       elseif expression 2
             commands evaluated if expression 2 is true
       elseif …

       .

       .

       .

       else
             commands evaluated if no other expression is true
       end

A.3   *Elementary matrices and matrix manipulation*

Elementary matrices and arrays

eye(n,n)	Identity matrix (its diagonal elements are 1 and all the others are 0)
linspace	Linspace($x1,x2$) generates 100 equally spaced points between $x1$ and $x2$
	Linspace($x1,x1,N$) generates N equally spaced points between $x1$ and $x2$
ones	ones(1,5) generates a row vector with its elements only ones
	ones(2,4) generates a 2×4 matrix with all its elements ones
rand	Uniformly distributed random numbers. $>>x = $ rand(1,5); x is a row vector of five elements of random numbers. $>> x = $ rand(2,3); x is a 2×3 matrix whose elements are random numbers
randn	Normally distributed random numbers. Applications are similar to rand above
zeros	Creates arrays and matrices of all zeros. $>>x = $ zeros(1,4); x is a row vector of four elements all with zero value. $>>x = $ zeros(3,4); x is a 3×4 matrix with all of its elements zero
: (colon)	Regularly spaced vector. $>>x = [1\ 4\ 2\ 5\ 8\ 3]; y = x(1,3{:}6);$ hence $y = [2\ 5\ 8\ 3]$
eps	Floating-point relative accuracy. To avoid NA response in case there exist a zero over zero expression at a point, as in the sinc function, we for example write $>>n = -4{:}4; x = \sin(n{*}pi{*}.1)./ \ldots$ $((n{*}pi + eps);$
i,j	Imaginary unit
pi	Ratio of a circle's circumference to its diameter

Matrix manipulation

diag	Diagonal matrices and diagonals of a matrix. $>>x = [1\ 3\ 5; \ldots$ $2\ 6\ 9; 4\ 7\ 0]; y = $ diag(x); will give a column vector $y = [1\ 6\ 0]^{T}$. $>>y1 = \ldots$ diag(x,1); will give a column vector $y1 = [3 \quad 9]^{T}$, which is the diagonal above the main diagonal
	$>>y2 = $ diag($x, -1$); will give the column vector $y2 = [2 \quad 7]$, which is the diagonal just below the main diagonal

$y3 = \text{diag}(\text{diag}(x))$; will give a 3×3 matrix with the diagonal 1, 6, 0 and the rest of the elements zero

fliplr	Flips vectors and matrices left–right
flipud	Flip matrices and vectors up–down
tril	Lower triangular part of a matrix including the main diagonal and the rest are zero. If $x = [1\ 3\ 5;\ 2\ 6\ 9;\ 4\ 7\ 0]$ then $y = \text{tril}(x)$ is the matrix $[1\ 0\ 0;\ 3\ 6\ 0;\ 4\ 7\ 0]$
triu	Upper triangular part of a matrix
toeplitz	Produces a Toeplitz matrix given a vector. $>>x = [1\ 5\ 2];\ y = \text{Toeplitz}(x)$ produces the matrix $y = [1\ 5\ 2;\ 5\ 1\ 5;\ 2\ 5\ 1]$

A.4   Elementary mathematics function

Elementary functions

abs	Absolute value of a number and the magnitude of a complex number
acos,acosh	Inverse cosine and inverse hyperbolic cosine
acot, acoth	Inverse cotangent and inverse hyperbolic cotangent
acsc,acsch	Inverse cosecant and inverse hyperbolic cosecant
angle	Phase angle of a complex number. $\text{angle}(1 + j) = 0.7854$
asec,asech	Inverse secant and inverse hyperbolic secant
asin,asinh	Inverse sine and inverse hyperbolic sine
atan,atanh	Inverse tangent and inverse hyperbolic tangent
ceil	Round toward infinity. For example, $\text{ceil}(4.22) = 5$
conj	Complex conjugate. $\text{conj}(2 + j*3) = 2 - j*3$
cos,cosh	Cosine and hyperbolic cosine
cot,coth	Cotangent and hyperbolic cotagent
csc,csch	Cosecant and hyperbolic cosecant
exp	Exponential. For example, $\exp(-1) = 1/e = 0.3679$
fix	Rounds toward zero. For example, $\text{fix}(-3.22) = -3$
floor	Round toward minus infinity. For example, $\text{floor}(-3.34) = -4$, and $\text{floor}(3.65) = 3$
imag	Imaginary part of a complex number. For example, $\text{imag}(2 + j*5) = 5$
log	Natural logarithm. For example, $\log(10) = 2.3026$
log 2	Based 2 logarithm. For example, $\log 2(10) = 3.3219$
log 10	Common (base 10) logarithm. For example, $\log 10(10) = 1$
mod	Modulus (signed remainder after division). For example, $\text{mod}(10,3) = 1$, $\text{mod}(10,4) = 2$. In general $\text{mod}(x,y) = x - n*y$
real	Real part of complex number
rem	Remainder after division. For example, $\text{rem}(10,3) = 1$, $\text{rem}(10,5) = 0$, $\text{rem}(10,4) = 2$
round	Round to the nearest integer. For example, $\text{round}(3.22) = 3$, $\text{round}(3.66) = 4$
sec,sech	Secant and hyperbolic secant

sign	Signum function. sign(x) = 0 for x = 0, sign(x) = 1 for $x > 0$ and sign(x) = −1 for $x < 1$
sin,sinh	Sine and hyperbolic sine
sqrt	Square root, e.g., sqrt(4) = 2
tan,tanh	Tangent and hyperbolic tangent
erf,erfc	Error and coerror function
gamma	Gamma function, e.g., gamma(6) = 120 or 1*2*3*4*(6 − 1) = 120

A.5   Numerical linear algebra

Matrix analysis

det   Matrix determinant >>a = [1 2; 3 4]; det(a) = 1 × 4 − 2 × 3 = −2

norm   The norm of a vector, e.g., norm(v) = sum(abs(v).^2)^(1/2)

rank   Rank of a matrix. Rank(A) provides the number of independent columns or rows of matrix A

trace   Sum of the diagonal elements, e.g., trace([1 3; 4 12]) = 13

eig   Eigenvalues and eigenvectors. >>[v,d] = eig([1 3; 5 8]); therefore,

$$v = \begin{bmatrix} -0.8675 & -0.3253 \\ 0.4974 & -0.9456 \end{bmatrix}, \quad d = \begin{bmatrix} -0.7202 & 0 \\ 0 & 9.7202 \end{bmatrix}$$

inv   Matrix inversion, e.g., >> A = [1 3; 5 8]; B = inv(A); therefore,

$$B = \begin{bmatrix} -1.1429 & 0.4286 \\ 0.7143 & -0.1429 \end{bmatrix}, \quad A*B = \begin{bmatrix} 1.0000 & 0 \\ 0 & 1.0000 \end{bmatrix}$$

A.6   Data analysis

Basic operations

max   Maximum element of an array. >>v = [1 3 5 2 1 7]; x = max(v); therefore, x = 7

mean   Average or mean value of an array, e.g., mean([1 3 5 2 8]) = 19/5 = 3.8

median   Median value of an array, e.g., median([1 3 5 2 8]) = 3

min   Minimum element of an array

sort   Sorts elements in ascending order, e.g., sort([1 3 5 2 8]) = [1 2 3 5 8]

std   Standard deviation

sum   Sum of an array elements, e.g., sum([1 3 5 2 8]) = 19

Filtering-convolution

conv   Convolution and polynomial multiplication, e.g., conv([1 1 1]) = [1 2 3 2 1], if we have to multiply these two polynomials (x^2 + 2x + 1)*(x + 2) we convolve their coefficients conv([1 2 1],[1 2]) = [1 4 5 2], therefore, we write the polynomial x^3 + 4x^2 + 5x + 2

conv2   2D convolution

filter   Filter data with infinite impulse response (IIR) or finite impulse
          response (FIR) filter. Let the FIR filter be given by $y(n) = 0.8x(n) +$
          $0.2x(n - 1) - 0.05x(n - 2)$. Let the input data are $x = [0.5 -0.2\ 0.6\ 0.1]$.
          Hence, $a = [1]$, $b = [0.5\ 0.2\ -0.05]$ and the output is given by $y =$
          filter(*a,b,x*). The result is: $y = [0.6250 -0.4063\ 0.8906 -0.1230]$.

Fourier transforms

abs   Absolute value and complex magnitude, e.g., abs(4 + j*3) = 5, abs
          ([−0.2 3.2]) = [0.2 3.2]
angle   Phase angle, e.g., angle(4 + j*3) = 0.6435 rad
fft   One-dimensional (1D) fast Fourier transform. >>$x = [1\ 1\ 1\ 0]$; $y =$
          fft(x);. Hence, $y = [3\ 0−1.0000i\ 1.0000\ 0 +1.0000i]$. If we had written
          $z = $fft($x$,8) we would have obtained $z = [3\ 1.7071\ −1.7071i\ 0\ −1.0000i$
          $0.2929 +0.2929i\ 1\ 0.2929 −0.2929i\ 0 +1.0000i\ 1.7071 +1.7071i]$
fft2   1D fast Fourier transform
fftshift Shift DC component of fast Fourier transform to center of spectrum.
          For example, we write in the command window: >>$x = [1\ 1\ 1\ 1\ 1\ ...$
          $0]$; $y = $fft($x$,256); then the command plot(abs(fftshift(y))) will center
          the spectrum. We can also write plot(abs(fftshift(fft(x,256))))
ifft   Inverse 1D fast Fourier transform
ifft2   Inverse 2D fast Fourier transform

A.7   *Two- and three-dimensional plotting*

Two-dimensional plots

plot   Linear plot. If we have three vectors of equal length such as x with
          numbers of equal distance, y and z, we can create the following
          simple plots: plot(y) will plot the values of y at numbers 1, 2, ... in
          the x-direction, plot(x,y) will plot the y values versus the equal-
          distance values of the vector x in the x-direction, plot(x,y,x,z) will
          plot both vectors y and z on the same graph, we can plot the two
          vectors by writing >>plot(x,y); hold on; plot(x,z); if we would like
          the second graph to have different color we write plot(x,z, "g") for
          green color
loglog   Log–log scale plot. For example, loglog(y) will produce the plot
semilogx Semilog scale plot. The log scale will be on the x-axis and the
          linear scale on the y-axis. The plot is accomplished by writing
          semilogx(y)
semilogy Semilog scale plot. The log scale will be on the y-axis and the
          linear scale on the x-axis. The plot is accomplished by writing
          semilogy(y)
axis   Controls axis scaling. For example, if we want the axes to have
          specific ranges, we write after we created a plot using the
          MATLAB default axis([minx maxx miny max])

grid   Grid lines. After we created the plot then we write grid on

subplot   Create axes in tiled positions. For example, when we write sub-plot(3,1,1) we expect 3×1 plots in one page starting plot one. Next, we write subplot(3,1,2) and then proceed to plot the second plot, etc. If we write subplot(3,2,1) we expect $3 \times 2 = 6$ plots on the page. After we write subplot(3,2,1) we proceed to plot the first of the 3×2 matrix format plots. For example, if we write subplot(3,2,2) and proceed to plot the figure we create a plot at line two and the second plot.

legend   Graph legend. For example, if we have two lines on the plot, one red and one green, and write legend("one", "two"), then a rect-angle frame will appear on the graph with a red line and the letters *one* and under a green line with the letters *two*

title   Graph title. For example, if we write title("This is a graph"), then the script in parenthesis will appear on the top of the graph

xlabel   X-axis label. For example, if we write xlabel("n time") the *n time* will appear under the *x*-axis

gtext   Place text with mouse. After we have created a plot, if we write in the command window gtext("this is the 1st graph") at the return a crosshair will appear on the graph and at the click the phrase in parenthesis will appear on the graph

Appendix B

Matrix analysis*

B.1 Definitions

Let A be an $m \times n$ matrix with elements a_{ij}, $i = 1, 2, \ldots, m$; $j = 1, 2, \ldots, n$. A shorthand description of A is

$$[A]_{ij} = a_{ij} \tag{B.1}$$

The **transpose** of A, denoted by A^T, is defined as the $n \times m$ matrix with elements a_{ji} or

$$[A^T]_{ij} = a_{ji} \tag{B.2}$$

Example B.1

$$A = \begin{bmatrix} 1 & 2 \\ 4 & 9 \\ 3 & 1 \end{bmatrix}, \quad A^T = \begin{bmatrix} 1 & 4 & 3 \\ 2 & 9 & 1 \end{bmatrix}$$

A **square** matrix is one for which $m = n$. A square matrix is **symmetric** if $A^T = A$.

The **rank** of a matrix is the number of linearly independent rows or columns, whichever is less. The **inverse** of a square $n \times m$ matrix A^{-1} for which

$$A^{-1}A = AA^{-1} = I \tag{B.3}$$

where I is the identity matrix

$$I = \begin{bmatrix} 1 & 0 & \cdots & 0 \\ 0 & 1 & \cdots & 0 \\ \vdots & \vdots & & \vdots \\ 0 & 0 & \cdots & 1 \end{bmatrix} \tag{B.4}$$

* In this appendix, uppercase letters represent matrices and lowercase letters without subscripts, excluding identifiers, indicate vectors. If lowercase letters have subscripts and indicate vectors will be written in bold-faced format.

265

A matrix A is **singular** if its inverse does not exist.

The determinant of a square $n \times m$ matrix is denoted by $\det\{A\}$, and it is computed as

$$\det\{A\} = \sum_{j=1}^{n} a_{ij} C_{ij}$$ (B.5)

where

$$C_{ij} = (-1)^{i+j} M_{ij}$$ (B.6)

and M_{ij} is the determinant of the submatrix A obtained by deleting the ith row and jth column and is called the **minor** of a_{ij}. C_{ij} is the **cofactor** of a_{ij}.

Example B.2

$$A = \begin{bmatrix} 1 & 2 & 4 \\ 4 & -3 & 9 \\ -1 & -1 & 6 \end{bmatrix}, \quad \det\{A\} = (-1)^{1+1} \begin{vmatrix} -3 & 9 \\ -1 & 6 \end{vmatrix} + (-1)^{1+2} 2 \begin{vmatrix} 4 & 9 \\ -1 & 6 \end{vmatrix} + (-1)^{1+3} 4 \begin{vmatrix} 4 & -3 \\ -1 & -1 \end{vmatrix}$$

$$= C_{11} + C_{12} + C_{13} = (18 + 9) + [-2(24 + 9)] + [4(-4 - 3)].$$

Any choice of i will yield the same value for the $\det\{A\}$.

A quadratic form Q is defined as

$$Q = \sum_{n=1}^{n} \sum_{j=1}^{n} a_{ij} x_i x_j$$ (B.7)

In defining the quadratic form, it is assumed that $a_{ji} = a_{ij}$. This entails no loss in generality since any quadratic functions may be expressed in this manner. Q may also be expressed as

$$Q = x^T A x$$ (B.8)

where $x = \begin{bmatrix} x_1 & x_2 & x_n \end{bmatrix}^T$ and A is a square $n \times m$ matrix with $a_{ji} = a_{ij}$ (symmetric matrix).

Example B.3

$$Q = \begin{bmatrix} x_1 & x_2 \end{bmatrix} \begin{bmatrix} a_{11} & a_{12} \\ a_{21} & a_{22} \end{bmatrix} \begin{bmatrix} x_1 \\ x_2 \end{bmatrix} = \begin{bmatrix} a_{11}x_1 + a_{21}x_2 & a_{12}x_1 + a_{22}x_2 \end{bmatrix} \begin{bmatrix} x_1 \\ x_2 \end{bmatrix}$$

$$= a_{11}x_1^2 + a_{21}x_1 x_2 + a_{12}x_1 x_2 + a_{22}x_2^2.$$

A square $n \times m$ matrix A is **positive semidefinite** if A is symmetric and

$$x^T A x \geq 0, \quad \text{for all } x \neq 0 \tag{B.9}$$

If the quadratic form is strictly positive, matrix A is called **positive definite**. If a matrix is positive definite or positive semidefinte it is automatically assumed that the matrix is symmetric.

The **trace** of a square matrix is the sum of the diagonal elements or

$$\text{tr}\{A\} = \sum_{i=1}^{n} a_{ii} \tag{B.10}$$

A partitioned $m \times n$ matrix A is one that is expressed in terms of its submatrices. An example is the 2×2 partitioned

$$A = \begin{bmatrix} A_{11} & A_{12} \\ A_{21} & A_{22} \end{bmatrix}, \quad \begin{bmatrix} k \times l & k \times (n-l) \\ (m-k) \times l & (m-k) \times (n-l) \end{bmatrix} \tag{B.11}$$

MATLAB® functions

```
B=A';% B is the transpose of A
B=inv(A);%B is the inverse of A
a=det(A);% a is the determinant of A
I=eye(n);%I is an nxn identity matrix
a=trace(A);%a is the trace of A
```

B.2   Special matrices

A **diagonal** matrix is a square $n \times n$ matrix with $a_{ij} \neq 0$ for $i \neq j$. A diagonal matrix has all the elements off the principal diagonal equal to zero. Hence

$$A = \begin{bmatrix} a_{11} & 0 & \cdots & 0 \\ 0 & a_{22} & \cdots & 0 \\ \vdots & \vdots & \vdots & \vdots \\ 0 & 0 & \cdots & a_{nn} \end{bmatrix} \tag{B.12}$$

$$A^{-1} = \begin{bmatrix} a_{11}^{-1} & 0 & \cdots & 0 \\ 0 & a_{22}^{-1} & \cdots & 0 \\ \vdots & \vdots & \vdots & \vdots \\ 0 & 0 & \cdots & a_{nn}^{-1} \end{bmatrix} \tag{B.13}$$

A generalization of the diagonal matrix is the square $n \times m$ block diagonal matrix

$$A = \begin{bmatrix} A_{11} & 0 & \cdots & 0 \\ 0 & A_{22} & \cdots & 0 \\ \vdots & \vdots & \vdots & \vdots \\ 0 & 0 & \cdots & A_{kk} \end{bmatrix} \tag{B.14}$$

where all A_{ii} matrices are square and the submatrices are identically zero. The submatrices may not have the same dimensions. For example, if $k = 2$, A_{11} may be a 2×2 matrix and A_{22} might be a scalar. If all A_{ii} are nonsingular, then

$$A^{-1} = \begin{bmatrix} A_{11}^{-1} & 0 & \cdots & 0 \\ 0 & A_{22}^{-1} & \cdots & 0 \\ \vdots & \vdots & \vdots & \vdots \\ 0 & 0 & \cdots & A_{kk}^{-1} \end{bmatrix} \tag{B.15}$$

and

$$\det\{A\} = \prod_{i=1}^{n} \det\{A_{ii}\} \tag{B.16}$$

A square $n \times n$ matrix is orthogonal if

$$A^{-1} = A^{\mathrm{T}} \tag{B.17}$$

Example B.4

$$A = \begin{bmatrix} \dfrac{2}{\sqrt{5}} & \dfrac{1}{\sqrt{5}} \\ -\dfrac{1}{\sqrt{5}} & \dfrac{2}{\sqrt{5}} \end{bmatrix}, \quad A^{-1} = \dfrac{1}{\det\{A\}} \begin{bmatrix} \dfrac{2}{\sqrt{5}} & \dfrac{1}{\sqrt{5}} \\ -\dfrac{1}{\sqrt{5}} & \dfrac{2}{\sqrt{5}} \end{bmatrix}^{\mathrm{T}} = \begin{bmatrix} \dfrac{2}{\sqrt{5}} & -\dfrac{1}{\sqrt{5}} \\ \dfrac{1}{\sqrt{5}} & \dfrac{2}{\sqrt{5}} \end{bmatrix} = A^{\mathrm{T}}$$

A matrix is **orthogonal** if its columns (and rows) are orthonormal. Therefore, we must have

$$A = [\mathbf{a}_1 \quad \mathbf{a}_2 \quad \cdots \quad \mathbf{a}_n]$$
$$\mathbf{a}_i^{\mathrm{T}} \mathbf{a}_j = \begin{cases} 0, & \text{for } i \neq j \\ 1, & \text{for } i = j \end{cases} \tag{B.18}$$

An idempotent matrix is a square $n \times n$ matrix which satisfies the relations

$$A^2 = A$$
$$A^m = A \tag{B.19}$$

Example B.5

The projection matrix $A = H(H^TH)^{-1}H^T$ becomes

$A^2 = H(H^TH)^{-1}H^T H(H^TH)^{-1}H^T = H(H^{-1}H^{-T}H^TH(H^TH)^{-1})H^T = H(H^{-1}IH(H^TH)^{-1})H^T = H(H^TH)^{-1}H^T$ and hence, it is an idempotent matrix.

A **Toeplitz** square matrix is defined as

$$[A]_{ij} = a_{i-j} \tag{B.20}$$

$$A = \begin{bmatrix} a_0 & a_{-1} & a_{-2} & \cdots & a_{-(n-1)} \\ a_1 & a_0 & a_{-1} & \cdots & a_{-(n-2)} \\ \vdots & \vdots & \vdots & \vdots & \vdots \\ a_{n-1} & a_{n-2} & a_{n-3} & \cdots & a_0 \end{bmatrix} \tag{B.21}$$

Each element along the northwest-to-southeast diagonals is the same. If in addition $a_{-k} = a_k$, then A is **symmetric Toeplitz**.

MATLAB functions

```
A = diag(x);%creates a diagonal matrix A with its
%diagonal the vector x;
A = toeplitz(x);%A is a symmetric Toeplitz matrix;
A = toeplitz(x,y)%x, and y must be of the same length,
%the main diagonal will
%be the first element of x, the first element of y is
%not used;
```

B.3   *Matrix operation and formulas*

Addition and subtraction

$$A + B = \begin{bmatrix} a_{11} & a_{12} & \cdots & a_{1n} \\ a_{21} & a_{22} & \cdots & a_{2n} \\ \vdots & \vdots & \vdots & \vdots \\ a_{m1} & a_{m2} & \cdots & a_{mn} \end{bmatrix} + \begin{bmatrix} b_{11} & b_{12} & \cdots & b_{1n} \\ b_{21} & b_{22} & \cdots & b_{2n} \\ \vdots & \vdots & \vdots & \vdots \\ b_{m1} & b_{m2} & \cdots & b_{mn} \end{bmatrix}$$
$$= \begin{bmatrix} a_{11}+b_{11} & a_{12}+b_{12} & \cdots & a_{1n}+b_{1n} \\ a_{21}+b_{21} & a_{22}+b_{22} & \cdots & a_{2n}+b_{2n} \\ \vdots & \vdots & \vdots & \vdots \\ a_{m1}+b_{m1} & a_{m2}+b_{m2} & \cdots & a_{mn}+b_{mn} \end{bmatrix} \tag{B.22}$$

Both matrices must have the same dimension.

Multiplication

$$AB(m \times n \times n \times k) = C(m \times k)$$

$$c_{ij} = \sum_{j=1}^{n} a_{ij} b_{ji} \tag{B.23}$$

Example B.6

$$AB = \begin{bmatrix} a_{11} & a_{12} \\ a_{21} & a_{22} \\ a_{31} & a_{32} \end{bmatrix} \begin{bmatrix} b_{11} & b_{12} \\ b_{21} & b_{22} \end{bmatrix} = \begin{bmatrix} a_{11}b_{11} + a_{12}b_{21} & a_{11}b_{12} + a_{12}b_{22} \\ a_{21}b_{11} + a_{22}b_{21} & a_{21}b_{12} + a_{22}b_{22} \\ a_{31}b_{11} + a_{32}b_{21} & a_{31}b_{12} + a_{32}b_{22} \end{bmatrix}, \quad 3 \times 2 \times 2 \times 2 = 3 \times 2 \tag{B.24}$$

Transposition

$$(AB)^{\mathrm{T}} = B^{\mathrm{T}} A^{\mathrm{T}} \tag{B.25}$$

Inversion

$$(A^{\mathrm{T}})^{-1} = (A^{-1})^{\mathrm{T}} \tag{B.26}$$

$$(AB)^{-1} = B^{-1} A^{-1} \tag{B.27}$$

$$A^{-1} = \frac{C^{\mathrm{T}}}{\det\{A\}}, \quad A \equiv n \times n \text{ matrix} \tag{B.28}$$

$$c_{ij} = (-1)^{i+j} M_{ij}, \quad M_{ij} \equiv \text{minor of } a_{ij} \text{ obtained by deleting the } i\text{th row}$$
and jth column of A \tag{B.29}

Example B.7

$$A^{-1} = \begin{bmatrix} 2 & 4 \\ -1 & 5 \end{bmatrix}^{-1} = \frac{1}{10+4} \begin{bmatrix} 5 & 1 \\ -4 & 2 \end{bmatrix}^{T} = \frac{1}{14} \begin{bmatrix} 5 & -4 \\ 1 & 2 \end{bmatrix},$$

$$AA^{-1} = \begin{bmatrix} 2 & 4 \\ -1 & 5 \end{bmatrix} \frac{1}{14} \begin{bmatrix} 5 & -4 \\ 2 & 2 \end{bmatrix} = \frac{1}{14} \begin{bmatrix} 14 & -8+8 \\ -5+5 & 4+10 \end{bmatrix} = \begin{bmatrix} 1 & 0 \\ 0 & 1 \end{bmatrix} = I$$

Determinant (see Equation B.5)

$A = n \times n$ matrix; $B = n \times n$ matrix

$$\det\{A^{\mathrm{T}}\} = \det\{A\} \tag{B.30}$$

$$\det\{cA\} = c^{n} \det\{A\} \tag{B.31}$$

$$\det\{AB\} = \det\{A\}\det\{B\} \tag{B.32}$$

$$\det\{A^{-1}\} = \frac{1}{\det\{A\}} \tag{B.33}$$

Trace (see Equation B.10)

$A = n \times n$ matrix; $B = n \times n$ matrix

$$\text{tr}\{AB\} = \text{tr}\{BA\} \tag{B.34}$$

$$\text{tr}\{A^{\mathsf{T}}B\} = \sum_{i=1}^{n}\sum_{j=1}^{n} a_{ij}b_{ij} \tag{B.35}$$

$$\text{tr}\{xy^{\mathsf{T}}\} = y^{\mathsf{T}}x, \quad x, y = \text{vectors} \tag{B.36}$$

Matrix inversion formula

$A = n \times n; B = n \times m; C = m \times m; D = m \times n;$

$$(A + BCD)^{-1} = A^{-1} - A^{-1}B(DA^{-1}B + C^{-1})^{-1}DA^{-1} \tag{B.37}$$

$$(A + xx^{\mathsf{T}})^{-1} = A^{-1} - \frac{A^{-1}xx^{\mathsf{T}}A^{-1}}{1 + x^{\mathsf{T}}A^{-1}x}, \quad x = n \times 1 \text{ vector} \tag{B.38}$$

Partition matrices

Examples of 2×2 partition matrices are given below.

$$AB = \begin{bmatrix} A_{11} & A_{12} \\ A_{21} & A_{22} \end{bmatrix}\begin{bmatrix} B_{11} & B_{12} \\ B_{21} & B_{22} \end{bmatrix} = \begin{bmatrix} A_{11}B_{11} + A_{12}B_{21} & A_{11}B_{2} + A_{12}B_{22} \\ A_{21}B_{11} + A_{22}B_{21} & A_{21}B_{12} + A_{22}B_{22} \end{bmatrix} \tag{B.39}$$

$$\begin{bmatrix} A_{11} & A_{12} \\ A_{21} & A_{22} \end{bmatrix}^{\mathsf{T}} = \begin{bmatrix} A_{11}^{\mathsf{T}} & A_{12}^{\mathsf{T}} \\ A_{21}^{\mathsf{T}} & A_{22}^{\mathsf{T}} \end{bmatrix} \tag{B.40}$$

$$A = \begin{bmatrix} A_{11} & A_{12} \\ A_{21} & A_{22} \end{bmatrix} = \begin{bmatrix} k \times k & k \times (n-k) \\ (n-k) \times k & (n-k) \times (n-k) \end{bmatrix}$$

$$A^{-1} = \begin{bmatrix} (A_{11} - (A_{12}A_{22}^{-1}A_{21})^{-1} & -(A_{11} - A_{12}A_{22}^{-1}A_{21})^{-1}A_{12}A_{22}^{-1} \\ -(A_{22} - A_{21}A_{11}^{-1}A_{12})^{-1}A_{21}A_{11}^{-1} & (A_{22} - A_{21}A_{11}^{-1}A_{12})^{-1} \end{bmatrix} \tag{B.41}$$

$$\det\{A\} = \det\{A_{12}\}\det\{A_{11} - A_{12}A_{22}^{-1}A_{21}\} = \det\{A_{11}\}\det\{A_{22} - A_{21}A_{11}^{-1}A_{12}\} \tag{B.42}$$

Important theorems

1. A square matrix A is singular (invertible) if and only if its columns (or rows) are linearly independent or, equivalently, if its det $\{A\} \neq 0$. If this is true A is of **full rank**. Otherwise it is singular.
2. A square matrix A is positive definite if and only if
   a.

$$A = CC^T \tag{B.43}$$

   where C is a square matrix of the same dimension as A and it is of full rank (invertible)
   b. The principal minors (the ith principal minor is the determinant of the submatrix formed by deleting all rows and columns with an index greater than i) are all positive. (The principal minor is the determinant of the submatrix formed by deleting all rows and columns with an index greater than i.) If A can be written as in Equation B.43, but C is not full rank or the principal minors are only nonnegative, then A is positive definite.
3. If A is positive definite, then

$$A^{-1} = (C^{-1})^T C^{-1} \tag{B.44}$$

4. If A is positive definite and B ($m \times n$) is of full rank ($m \leq n$) then BAB^T is positive definite.
5. If A is positive definite (or positive semidefinite), then the diagonal elements are positive (nonnegative).

B.4   *Eigendecomposition of matrices*

Let λ denotes an **eigenvalue** of the matrix A ($n \times n$), then

$$Av = \lambda v \tag{B.45}$$

where v is the eigenvector corresponding to the eigenvalue λ. If A is symmetric, then

$$Av_i = \lambda_i v_i, \quad Av_j = \lambda_j v_j, \quad \lambda_i \neq \lambda_j$$

and

$$v_j^T A v_i = \lambda_i v_j^T v_i \quad \text{(a)}$$

$$v_i^T A v_j = \lambda_j v_i^T v_j \quad \text{or} \quad v_j^T A v_i = \lambda_i v_j^T v_i \quad \text{(b)}$$

Subtracting (a) from (b), we obtain $(\lambda_i - \lambda_j) v_i^T v_i = 0$. But $\lambda_i \neq \lambda_j$ and hence $v_j^T v_i = 0$ which implies that the eigenvectors of a symmetric matrix are orthogonal. We can proceed and normalize them producing orthonormal eigenvectors.

From Equation B.45 we write

$$A\begin{bmatrix} v_1 & v_2 & \cdots & v_n \end{bmatrix} = \begin{bmatrix} \lambda_1 v_1 & \lambda_2 v_2 & \cdots & \lambda_n v_n \end{bmatrix} \quad \text{or} \quad AV = V\Lambda \quad \text{(B.46)}$$

where

$$\Lambda = \begin{bmatrix} \lambda_1 & 0 & 0 & \cdots & 0 \\ 0 & \lambda_2 & 0 & \cdots & 0 \\ 0 & 0 & \lambda_3 & \cdots & 0 \\ \vdots & \vdots & \vdots & \vdots & \vdots \\ 0 & 0 & 0 & \cdots & \lambda_n \end{bmatrix} \quad \text{(B.47)}$$

Because v_i are mutually orthogonal, $v_i^T v_j = \delta_{ij}$ makes V a **unitary** matrix, $V^T V = I = VV^T$. Postmultiply Equation B.46 by V^T, we obtain

$$A = V\Lambda V^T \sum_{i=1}^{n} \lambda_i v_i v_i^T \quad \text{(B.48)}$$

which is known as **unitary decomposition** of A. We also say that A is **unitary similar** to the diagonal Λ, because a unitary matrix V takes A to diagonal form: $V^T A V = \Lambda$.

If $\Lambda = I$, then from Equation B.48 $A = VV^T = I$ and, hence,

$$I = VV^T = \sum_{i=1}^{n} v_i v_i^T \quad \text{(B.49)}$$

Each of the terms in the summation is of rank 1 projection matrix:

$$P_i^2 = v_i v_i^T v_i v_i^T = v_i v_i^T = P_i, \quad v_i^T v_i = 1 \quad \text{(B.50)}$$

$$P_i^T = v_i v_i^T = P_i \quad \text{(B.51)}$$

Hence, we write (see Equations B.48 and B.49)

$$A = \sum_{i=1}^{n} \lambda_i P_i \quad \text{(B.52)}$$

$$I = \sum_{i=1}^{n} P_i \tag{B.53}$$

Inverse

Because V is unitary matrix, $VV^T = I$ or $V^T = V^{-1}$ or $V = (V^T)^{-1}$ and, therefore,

$$A^{-1} = (V^T)^{-1} \Lambda^{-1} V^{-1} = V \Lambda^{-1} V^T = \sum_{i=1}^{n} \frac{1}{\lambda_i} v_i v_i^T \tag{B.54}$$

Determinant

$$\det\{A\} = \det\{V\} \det\{\Lambda\} \det\{V^{-1}\} = \det\{\Lambda\} = \prod_{i=1}^{n} \lambda_i \tag{B.55}$$

B.5   Matrix expectations

$$E\{x\} = m_x \tag{B.56}$$

$$E\{(x - m_x)(x - m_x)^T\} = R_{xx} \tag{B.57}$$

$$E\{\text{tr}\{A\}\} = \text{tr}\{E\{A\}\} \tag{B.58}$$

$$E\{Ax + b\} = Am_x + b \tag{B.59}$$

$$E\{xx^T\} = R_{xx} + m_x m_x^T \tag{B.60}$$

$$E\{xa^T x\} = (R_{xx} + m_x m_x^T)a \tag{B.61}$$

$$E\{(x + a)(x + a)^T\} = R_{xx} + (m_x + a)(m_x + a)^T \tag{B.62}$$

$$E\{x^T x\} = \text{tr}\{R_{xx}\} + m_x^T m_x = \text{tr}\{R_{xx} + m_x m_x^T\} \tag{B.63}$$

$$E\{x^T a x^T\} = a^T \{R_{xx} + m_x m_x^T\} \tag{B.64}$$

$$E\{x^T A x\} = \text{tr}\{A R_{xx}\} + m_x^T A m_x = \text{tr}\{A(R_{xx} + m_x m_x^T)\} \tag{B.65}$$

B.6 Differentiation of a scalar function with respect to a vector

$$x = \begin{bmatrix} x_1 & x_2 & \cdots & x_n \end{bmatrix}^{\mathrm{T}}, \quad \frac{\partial}{\partial x} = \begin{bmatrix} \frac{\partial}{\partial x_1} & \frac{\partial}{\partial x_2} & \cdots & \frac{\partial}{\partial x_n} \end{bmatrix}^{\mathrm{T}}$$

$$\frac{\partial}{\partial x}(y^{\mathrm{T}}x) = \frac{\partial}{\partial x}(x^{\mathrm{T}}y) = y \tag{B.66}$$

$$\frac{\partial}{\partial x}(x^{\mathrm{T}}A) = A \tag{B.67}$$

$$\frac{\partial}{\partial x}(x^{\mathrm{T}}) = I \tag{B.68}$$

$$\frac{\partial}{\partial x}(x^{\mathrm{T}}x) = 2x \tag{B.69}$$

$$\frac{\partial}{\partial x}(x^{\mathrm{T}}Ay) = Ay \tag{B.70}$$

$$\frac{\partial}{\partial x}(y^{\mathrm{T}}Ax) = A^{\mathrm{T}}y \tag{B.71}$$

$$\frac{\partial}{\partial x}(x^{\mathrm{T}}Ax) = (A + A^{\mathrm{T}})x \tag{B.72}$$

$$\frac{\partial}{\partial x}(x^{\mathrm{T}}Ax) = 2Ax, \quad \text{if } A \text{ is symmetric} \tag{B.73}$$

$$\frac{\partial}{\partial x}(a^{\mathrm{T}}Axx^{\mathrm{T}}) = (A + A^{\mathrm{T}})xx^{\mathrm{T}} + x^{\mathrm{T}}AxI \tag{B.74}$$

Appendix C

Lagrange multiplier method

To solve a constrained optimization problem, the Lagrange method is used. The normalized least mean square (NLMS) recursion, for example, can be obtained as a solution to the following problem:

minimize $\min_{\mathbf{w}} \| \mathbf{w}(n) - \mathbf{w}(n-1)^2 \|$ subject to constraint $d(n) = \mathbf{w}^{\mathrm{T}}(n+1)\mathbf{x}(n)$

The first step in the solution is to write the cost function as follows:

$$J(n) = [\mathbf{w}(n+1) - \mathbf{w}(n)]^{\mathrm{T}}[\mathbf{w}(n+1) - \mathbf{w}(n)] + \lambda[d(n) - \mathbf{w}^{\mathrm{T}}(n+1)\mathbf{x}(n)] \quad \text{(C.1)}$$

Differentiating the cost function above with respect to $\mathbf{w}(n + 1)$, we obtain

$$\frac{J(n)}{\partial \mathbf{w}(n+1)} = 2[\mathbf{w}(n+1) - \mathbf{w}(n)] - \lambda \mathbf{x}(n) \quad \text{(C.2)}$$

Setting the results to zero results in

$$\mathbf{w}(n+1) = \mathbf{w}(n) + \frac{1}{2}\lambda \mathbf{x}(n) \quad \text{(C.3)}$$

Substituting this result into the constraint $d(n) = \mathbf{w}^{\mathrm{T}}(n + 1)\mathbf{x}(n)$, we obtain

$$d(n) = \left(\mathbf{w}(n) + \frac{1}{2}\lambda \mathbf{x}(n)\right)^{\mathrm{T}}, \quad \mathbf{x}(n) = \mathbf{w}^{\mathrm{T}}(n)\mathbf{x}(n) + \frac{1}{2}\lambda \| \mathbf{x}(n) \|^2 \quad \text{(C.4)}$$

Since $e(n) = d(n) - \mathbf{w}^{\mathrm{T}}(n)\mathbf{x}(n)$, solving Equation C.4 for λ leads to

$$\lambda = \frac{2e(n)}{\| \mathbf{x}(n) \|^2} \quad \text{(C.5)}$$

Substituting Equation C.5 in Equation C.3, we find

$$\mathbf{w}(n+1) = \mathbf{w}(n) + \frac{1}{\| \mathbf{x}(n) \|^2} e(n)\mathbf{x}(n) \quad \text{(C.6)}$$

Finally, introducing a factor μ in Equation C.6 to control the change in the weight vector, we obtain the conventional NLMS algorithm

$$\mathbf{w}(n+1) = \mathbf{w}(n) + \frac{\mu}{||\mathbf{x}(n)||^2} e(n)\mathbf{x}(n) \qquad (C.7)$$

Index